Modularity Methodology

Peter Adekunle · Clinton Aigbavboa · Opeoluwa Akinradewo

Modularity Methodology

Springer

Peter Adekunle
Department of Construction Management
and Quantity Surveying
University of Johannesburg
Johannesburg, Gauteng, South Africa

Clinton Aigbavboa
Department of Construction Management
and Quantity Surveying
University of Johannesburg
Johannesburg, Gauteng, South Africa

Opeoluwa Akinradewo
Department of Quantity Surveying
and Construction Management
University of Free State
Bloemfontein, Free State, South Africa

ISBN 978-3-032-12917-8 ISBN 978-3-032-12918-5 (eBook)
https://doi.org/10.1007/978-3-032-12918-5

This Springer imprint is published by the registered company Springer Nature Switzerland AG
The registered company address is: Gewerbestrasse 11, 6330 Cham, Switzerland

Competing Interests The authors have no competing interests to declare that are relevant to the content of this manuscript.

Contents

Part I
Background Information

Chapter 1
Introduction to Modularity Methodology in the Construction Industry

Abstract This chapter provides a basic understanding of modular construction and its increasing importance in contemporary project delivery by introducing the idea of modularity methodology within the framework of the construction industry. Buildings that are constructed off-site, under controlled plant conditions, with the same materials and designed in accordance with the same codes and standards as conventionally built facilities, but in roughly half the time, are referred to as modular construction. Because of its potential to shorten construction times, cut waste, improve quality control, and increase safety, this strategy is becoming increasingly popular. The chapter opens with a summary of the global issues that call for more economical, sustainable, and efficient building techniques, presenting modular construction as a workable answer. The problem statement highlights the shortcomings of conventional building techniques in meeting modern demands like project delays, labour shortages, urbanisation, and environmental concerns. The chapter defines modular construction and explains its basic concepts, including prefabrication, standardisation, and off-site production. A historical overview traces the development of modular building from its first applications in post-war housing to its current applications in commercial, residential, and infrastructure projects. The book aims to provide comprehensive insights into modular construction techniques, their benefits, drawbacks, and the factors influencing their adoption in different geographic areas. The chapter also summarises the book's primary objectives and contributions. Additionally, it discusses how modularity in construction promotes technological innovation, sustainability, and productivity. The book's format serves as a roadmap for the reader as they progress through the following chapters, which cover case studies, theoretical frameworks, best practices, and new trends.

1.1 Introduction

Construction is the process of putting together various elements according to a specified design. Construction methods are the procedures that professionals use to erect buildings such as houses, offices, and other structures (Kineber et al., 2023).

P. Adekunle et al., *Modularity Methodology*,
https://doi.org/10.1007/978-3-032-12918-5_1

Costs, available resources, the team's degree of expertise, and the project's location frequently impact the building approach a team of specialists choose to use. Understanding construction practices is essential for architects, engineers, and construction professionals (Gal et al., 2008). This knowledge aids in designing safe workplaces, project planning, and building structures. Furthermore, it is advantageous to be aware of this information when submitting applications for positions in the construction sector. Professionals widely use the building structure, primary usage, and materials for the building's foundation and frame to identify different construction techniques (Zhou et al., 2012). The choice of building method can significantly influence the skill set required for construction. For instance, specific methods may demand specialised skills such as steel frame welding, advanced concrete pouring techniques, or the precise assembly of prefabricated components. Typically, the building's project manager or architect selects the construction method, ensuring it aligns with the project's goals, budget, and timeline (Saad et al., 2023). However, clients may specify a preferred construction method based on their needs, preferences, or previous experiences. This selection process is critical as it impacts the construction project's efficiency, cost, and sustainability.

Regona et al. (2022) defined the construction process as the series of precise steps necessary to complete a building project, broadly categorised into five phases: planning/design, pre-construction, procurement, construction, and post-construction. Each phase presents unique challenges, influenced by the project's scale, the number of stakeholders involved, budget constraints, and deadlines. During the planning and design phase, architects and engineers develop comprehensive blueprints that outline the project's specifications and requirements (Pizarro & Massone, 2021). This stage is crucial for establishing the project's feasibility and setting the foundation for successful execution. The pre-construction phase involves detailed planning and preparation, including securing permits, finalising designs, and setting up the project site (Ndekugri et al., 2022). This phase requires meticulous coordination to ensure all necessary resources are available and potential obstacles are identified and mitigated. Procurement, the third phase, focuses on sourcing the materials, labour, and equipment required for construction. Effective procurement strategies are essential to avoid delays and cost overruns, as the timely delivery of high-quality materials directly impacts the project's progression.

Baduge et al. (2022) opined that the construction phase is where the actual building takes place, translating plans into physical structures. This stage is typically resource-intensive and time-consuming, involving many tasks and coordination among various trades and professionals. Challenges here include labour, ensuring safety (Amoah & Simpeh, 2021), adhering to timelines (Adekunle et al., 2022), and maintaining quality standards. Despite thorough planning, unforeseen weather conditions, labour shortages, and logistical problems often arise, requiring adaptive management and problem-solving skills. Post-construction, the final phase, encompasses activities such as inspection, testing, and commissioning of the building to ensure it meets all specified criteria and regulatory standards (Chau et al., 2015). This phase also involves addressing defects, completing the final touches, and officially handing the project to the client. Additionally, the post-construction phase may include training the building's operators and maintenance staff to ensure smooth operation.

Furthermore, McLeod et al. (2012)) highlighted that regardless of a project's size, construction projects are inherently complex and prolonged, with many interconnected components and stakeholders. The intricacies of managing these projects demand high levels of coordination, communication, and strategic planning. Russell-Smith et al. (2015) observed an increased focus on the operational phase of buildings, particularly concerning energy use and indoor air quality. This shift reflects a growing awareness of the importance of sustainable building operations in promoting occupant health and well-being. Modern buildings are now designed with advanced environmental plans incorporating high-performance features to enhance energy efficiency and ensure a healthy indoor environment. As such, considerations of building operation have become integral to the overall construction process, emphasising the need for continuous innovation and adherence to best practices in sustainable building design and construction.

The eighteenth century saw the start of the Industrial Revolution, which resulted in significant technical improvements (Rymarczyk, 2020). However, the development of building materials made significant strides during the nineteenth century. Cast iron, wrought iron, and later steel allowed for new structures such as railroads, bridges, and building frames (Krishna, 2021). Glass was also used for steel-framed buildings with large, glazed envelopes. It seems that little innovation occurred in construction during this time. Groumpos (2020) reported that later in the nineteenth century, a new industrial sector that produced building equipment (including elevators, boilers, radiators, pipes, and sanitary appliances) began to emerge. The Industrial Revolution was pivotal in advancing construction methods by significantly increasing the availability of building materials such as steel and concrete (Wuyts et al., 2019). This period marked the beginning of more efficient and large-scale production processes, enabling the construction of more robust and durable structures.

The demand for new dwellings surged during World Wars I and II, especially the latter, as many cities were heavily damaged and required extensive rebuilding efforts. The industrialisation of buildings results from the need to switch from antiquated, labour-intensive processes to more modern ones. In the construction sector, industrialisation is defined in a variety of ways. George and George (2020) highlighted that industrialisation is characterised by the construction process rather than the result, in addition to being defined as mechanisation. The use of modern framework (such as 'tunnel' shutters) and pre-stressing methods, as well as the introduction of new technologies such as prefabrication or modern in situ processes such as the various applications of slip-forms for chimneys, bunkers, and silos, are all part of the industrialisation of the building industry, according to Ngowi et al. (2005). Contemporary design methodologies informed by scientific knowledge of building physics, fire science, and computer technology are additional indicators of industrialisation. One sort of industrialisation is modularisation, which is based on the industrial manufacturing of construction components off-site or nearby.

Modularisation is currently achievable with all structural materials, including concrete, steel, aluminium, and polymers (Kolbeck et al., 2023). Even pre-cutting of stones has been employed to facilitate modular construction. Various terms, such as portal homes, mobile homes, manufactured homes, and system buildings, refer to multi-story industrial buildings and dwellings produced using prefabrication (Giorgio et al., 2022). Prefabrication involves the off-site manufacturing of building components, which are transported and assembled on-site. This method offers numerous advantages, including reduced construction time, improved quality control, and minimised waste, making it a popular choice in modern construction. With the development of concrete material, modularisation acquired fresh traction for civil engineering projects such as bridges and residences. Russell (2012) informs that the building systems method is of limited use when conceived as a panacea by an architect, administrator, or politician. However, the technique has also received criticism. Unlike in many other countries such as the USA, France, or Scandinavia, prefabrication lost appeal in the UK. Despite this, modularisation has often been employed on sites with complex local characteristics, aiding in sustainable development. The adaptability of modular construction to various challenging environments makes it a valuable approach for creating efficient and sustainable buildings in areas where traditional construction methods may be less feasible.

Nevertheless, owing to growing awareness of environmental contamination, the depletion of natural resources, and the societal issues they raise, sustainable development and building are becoming increasingly essential topics on a global scale (Wanner, 2015). Buildings are one of the biggest consumers of natural resources and a significant source of greenhouse gas emissions. In the US, buildings account for 38.9% of primary energy use, 38% of total carbon dioxide emissions, and 30% of waste generation (Onat et al., 2014). These statistics highlight buildings' significant impact on energy consumption, greenhouse gas emissions, and waste generation. As a result, improving the energy efficiency and sustainability of buildings is crucial for mitigating climate change and reducing environmental impacts. Traditional on-site building methods have a long history of receiving flak for their low output, poor quality and safety records, lengthy construction schedules, and significant garbage creation. Modularisation, a manufacturing technique in which several materials are joined to produce a component element of the final installation, mitigates the issues caused by traditional building methods (Adekunle et al., 2024a, 2024b, 2024c). Numerous benefits of using modular technology in construction have been extensively discussed in earlier literature (Lawson et al., 2012; Thai et al.,2020; Tsz Wai et al., 2023). These benefits include quicker construction, cost optimisation, better quality, increased durability, better architectural appearance, improved occupational health and safety, material conservation, decreased waste at construction sites, lower environmental emissions, and lower energy and water use. Modularisation has a greater likelihood of helping sustainable building projects because of these advantages.

In 1996, Germany (31%), Sweden (41%), the Netherlands (40%), and Denmark (43%) had the highest precast levels globally (Zhang et al., 2014). Just 6% of reinforced concrete buildings in the United States are supplied by precast businesses, compared to an average of 18% for the European Union. Modular concrete systems only account for about 1.2% of the building construction market, even though the American modular concrete industry produces technologically and aesthetically complex buildings and building elements such as double tees, hollow-core slab elements, inverted tee and ledger beams, and façade panels (Ghaffar et al., 2018). Therefore, addressing modular issues in concrete structures is more important to achieve sustainable construction in developed and developing countries such as the United States, England, South Africa, and Kenya. It is essential to address significant challenges in the construction industry by promoting sustainable construction via modularity methodology. These challenges may include transportation logistics (Shin et al., 2022), integration with on-site construction (Jiang et al., 2022), adaptation of building codes (Davila Delgado et al., 2019), and awareness and acceptance (Xu et al., 2020) within the construction industry.

Developed and developing countries can encourage sustainable construction techniques, lessen their negative effects on the environment, and promote robust and energy-efficient structures by investing in research, modularity methodology, and development of construction methodology. This research project will produce a model for applying modularity methodology in the construction industry. Researchers have identified efficiency, affordable housing, job creation, skill development and disaster resilience as key attributes of modularity methodology to the construction industry. However, as stated earlier, there is no evidence of a modularity methodology for developing countries. It is not certain that characteristics that formulate modular methodologies in developed countries will also be applicable in developing countries. Therefore, this research is crucial to determine the feasibility and adaptation of modularity methodology in such contexts.

1.2 Problem Statement

The construction sector has been lagging behind other industries in productivity for a long time. Modular construction allows the sector to transition dramatically by relocating several building-related activities from traditional construction sites to factories with off-site, manufacturing-style operations (Adekunle et al., 2024a). Recent technological advancements, economic constraints, and evolving cultural perspectives have sparked a remarkable interest and investment in modular (or prefabricated) buildings. This trend could substantially enhance industrial output, alleviate housing shortages in numerous areas, and fundamentally revolutionise current construction practices if adopted broadly. Modular construction often comprises creating standardised building components in a factory off-site before assembling them on-site (Boyd et al., (2013). The words 'off-site construction,' 'prefabrication', and 'modular construction' are all interchangeable. These terms include various techniques and arrangements, from straightforward parts that are joined and interfaced to 3-D volumetric units with full fixtures. Modular buildings were financially viable in the eighteenth century, especially between 1848 and 1855, but the widespread adoption was fleeting (MacDuffie, 2013). This was due to the transient nature of gold mining communities and the economic realities of the time. Many of the people who arrived were in search of quick fortunes and thus were highly mobile, moving from one location to another as they followed gold leads. This meant that the demand for permanent, stable housing was limited. Moreover, the construction technologies and materials used in modular construction at that time were relatively rudimentary compared to today's standards (Adekunle et al., 2024b). The quality and durability of these early modular homes might not have met the long-term needs of a more permanent settlement.

Recent research, such as that of Khadodra et al. (2020) and Otasowie et al. (2024) indicates that modular buildings now possess enduring force. Technological advancements, cost efficiency, time efficiency, sustainability, adaptability, flexibility, and improved quality control are key factors establishing the enduring force of modern modular construction. As a result, there has recently been a surge in investment and interest. For instance, introducing digital tools has radically changed the modular construction idea by expediting supply logistics and simplifying module design (Bertram et al., 2019). Prefabricated home attitudes are beginning to change significantly as new, more intriguing material alternatives improve the aesthetic appeal of the buildings. To fully enjoy the benefits of modular construction, various stakeholders must make various decisions. On the other hand, investors need to focus more on markets susceptible to disruption and trends that identify winners and losers and provide savings. For instance, developers must establish portfolio divisions that might act as their "product core" (Cusumano et al., 2015, p. 11). Cost-cutting actions can be conducted in a variety of contexts. The use of modular construction firstly does away with the requirement for subcontractors and the margins they add to their quotes. The following significant trade-offs weigh prospective increases in material prices and an increase in logistics costs against decreases in on-site labour. However, compared to the costs for maintenance and redesign, the original design expenditures for modular projects are often higher (Veilleux et al., 2020). Considering these trade-offs, the projects most likely to achieve the most significant cost savings are those with the highest proportion of labour-intensive tasks and the highest levels of repeatability.

Furthermore, there has been a decline in construction standards, and the cost of owning a building continues to escalate. As a result, there is a critical need for a new, innovative system to address these challenges in the construction industry (Wuni & Shen, 2020). The study introduces an approach that leverages modular deployment in construction processes and products to improve product quality and process efficiency. The primary goal is optimising resources and the workforce to deliver high-quality products. Studies have extensively explored the concept of modularity across various fields (Campagnolo & Camuffo, 2010; Gershenson et al., 2004; Sievers et al., 2016), highlighting economies of scale as a significant benefit of modularity. Economies of scale refer to the ability to increase production output significantly while only incurring marginal cost increases (Şeker & Aydinli, 2023). This efficiency gain occurs because fixed costs, such as equipment and facility expenses, are spread over a more significant number of units produced. As production scales up, variable costs also decrease owing to the bulk purchasing of materials and more efficient use of labour. This principle aligns with the foundational capacities of production lines, designed to maximise output and minimise costs through streamlined processes and standardised components. The process of modularity is well understood, and its advantages are broadly recognised across multiple industries (Akinradewo et al., 2023). However, a dearth of research focuses on modularity methodology within the construction sector, particularly in contexts of developing countries construction industry. This gap underscores the importance of the current study.

The construction industry plays a vital role in the growth and performance of economies worldwide, contributing significantly to developed and developing nations' gross domestic product (GDP) (Pedercini et al., 2018). Its impact extends beyond mere economic metrics, influencing various societal development and infrastructure provision aspects. However, despite its importance, the industry is characterised by slow growth, low productivity, and safety and material waste management concerns. These issues have prompted extensive research and initiatives to transform the industry through lean methodology, prefabrication, and other construction innovations worldwide. However, there has been insufficient focus on applying modular methodological systems in the construction industry, particularly in developing countries (Otasowie et al.,2025).

Studying modularity methodology in this context is essential to address these challenges and foster significant improvements in efficiency, safety, and sustainability within the construction sector. Previous models of modularity methodology established in developed countries cannot be directly applied in developing countries, and the determinants of modularity methodology that can accelerate the construction process and enhance overall project efficiency in these regions are rarely known from past research. Therefore, it is essential to conduct new research focused on developing countries. Construction experts and stakeholders can simplify projects, minimise mistakes and delays, and generate more advantageous and economical results for all parties engaged in building processes by creating a modularity methodology model tailored to these situations.

1.3 Definition of Modular Construction

Modular construction takes place when a building is constructed off-site, under controlled plant conditions, using the same materials and designed to the same codes and standards as conventionally built facilities, but in about half the time (Murali & Sambath, 2020). Buildings are produced in 'modules' that, when put together on site, reflect the identical design intent and specifications of the most sophisticated site-built facility without compromise. Modular construction is sometimes referred to as volumetric construction, prefabricated buildings, or offsite construction, although strictly speaking, a prefabricated building need not be modular. Offsite construction encompasses a wide range of construction methods and techniques. In contrast, modular construction is a subset of off-site construction involving building entire modules off-site and assembling them on the final construction site.

Buildings can be constructed using modular 'parts' such as walls, frames, doors, ceilings, windows, or several 'complete' prefabricated modular building units. The modular building became popular after the Second World War when there was a need for the rapid construction of buildings (in particular dwellings) to replace bomb-damaged buildings and to accommodate returning troops (Baghchesaraei et al., 2016). They were initially well-received, but as they often remained in use well beyond their design life (for example, 'temporary' classrooms, some of which are still in use) and were sometimes aesthetically less attractive than traditional buildings, they fell out of favour. More recently, there has been a resurgence of interest in modular buildings, with a recognition that they can be aesthetically pleasing, achieve a very high build quality and be both long-lasting and sustainable (Armstrong, 2023). A new fleet of modular buildings has emerged that includes a wide variety of sizes and configurations, with sector-specific building types and styles and options for variable plans and multi-storey configurations.

Additionally, any modular building or component designed and constructed in compliance with this standard, and that is either entirely or largely made, fabricated, formed, or assembled in manufacturing plants, is intended for installation or on-site assembly at a building location (Tuvayanond & Prasittisopin, 2023). This has been manufactured so that all parts or processes cannot be inspected at the installation site without disassembly, damage, or destruction. Modular construction is sometimes referred to as 'prefab' in some regions. It refers to preassembling or prefabricating parts of the building at a second work site (i.e. the factory). One of the vital factors achieved through the adoption of modular construction is that it offers an accelerated schedule (Chen et al., 2021a, 2021b). While modules are built in controlled offsite locations, site work can co-occur. As illustrated above, parallel construction work can be performed in modular construction. Wrigley et al. (2021) observed that this modular process allows reduced construction times of about 30–50% that of conventional construction. In modular construction, floors, walls, ceilings, rafters, and roofs can be constructed simultaneously. On the other hand, during conventional construction, walls cannot be set until floors are in position; additional floors, ceilings and rafters cannot be added until walls are erected.

Modular construction makes it possible to optimise the purchase and usage of construction materials while minimising onsite waste and offering a higher-quality product to the buyer (Tam et al., 2015). According to the UK Group of Waste and Resources Action Programmes (WRAP), up to a 90% reduction in waste can be achieved through modular construction (Gálvez-Martos et al., 2018). Minimised materials waste includes wood pallets, shrink wrap, cardboard, plasterboard, timber, concrete, bricks, and cement. The modular structure is constructed off-site simultaneously with the foundation and other site work, reducing the time and impact on the surrounding site environment and the number of vehicles and equipment needed. When the owner's needs change, some modular buildings can be disassembled, relocated, or refurbished for their subsequent use, thus reducing the demand for raw materials and minimising the energy expended to create a building to meet the new need. Modular buildings can also contribute to leadership in energy and environmental design (LEED) requirements in the same categories as traditional, site-built construction (Chen et al., 2010). It can even provide an advantage regarding sustainable sites, energy and atmosphere, materials and resources, and indoor environmental quality. LEED certification has brought green buildings into the mainstream. It is the most widely recognised building sustainability credential architects, project developers, homeowners, and interior designers can receive. According to Ilhan and Yaman 2016), modular construction can also provide an advantage in similar categories in the International Green Construction Code (IGCC).

Furthermore, modular construction is process-oriented construction, which lends itself to being more efficient. Modular buildings are built with all the same materials and to the same building codes and architectural specifications as traditional onsite construction but in an indoor quality-controlled environment (Navaratnam et al., 2019). A prefabricated building module has built-in quality checks such as a car assembly line. Modular building factories are dedicated to crafting highly efficient buildings with inspections at each station, eliminating variability in decisions or unexpected complications that can occur in the field. Because modular units must be trucked long distances and then moved by crane or other method onto a foundation, they must also be highly structurally sound. Hence, modular buildings can be more durable than structures built onsite.

Modular construction sites come in two formats permanent facilities and temporary flying factories (Rosarius & Garcia de Soto, 2021). Permanent facilities service all construction sites in each area. Freight shipping can become expensive for projects at the edge of the service area; nevertheless, it still ends up being a more cost-effective and time-saving option than traditional building techniques. Flying factories are temporary facilities that provide components to a construction site before moving to the following location. This option can be more cost-effective for companies that operate in a broad area. Kamali & Hewage (2016) further stated that modular construction has the added benefit of setting up close to or on a building site, reducing or eliminating transportation costs and emissions. Having these locations closer to the source of necessary raw material components is also possible to minimise transportation costs while streamlining the supply chain.

Structural building components are often mistaken for off-site or modular construction. Structural building components are structures designed to carry a load or support weight. They can be wall panels, roof and floor trusses, columns and pillars, foundations, plinths, and even door and window frames. In traditional construction, all these components are built and assembled on site. Modular construction allows some of these structural components to be manufactured elsewhere, saving time and money (Islam et al., 2016). It is important to note that while some structural building components may be utilised in factory-built modular or mobile home construction, they are entirely separate from off-site construction. The manufacturing steps for structural components necessary for modular construction are similar to traditional methods, as stated by Pervez et al. (2021). Instead of gathering the basic materials at the job site and building everything from scratch, companies can manufacture each structural component in a factory setting.

1.4 History and Development of Modular Construction

According to Bendi et al. (2021), the history of modular buildings started in 1837. The first on-record example of this type of housing was a home designed and constructed by London-based carpenter Henry Manning. Upon creation, it was transported to Australia. Many more were built from this owing to its popularity and industry growth. Manning created the building for his son, who was due to emigrate to Australia (Agha et al., 2021). The appearance of his unique and innovative home thrilled the locals. Manning's design had caught on. The design is presumed to interest people owing to its easy transportation and assembly. By 1853, hundreds of these modular builds had been shipped to Australia yearly. However, it was not just Australia embracing modular buildings; they were soon sighted all over the globe.

Also, modular buildings were spotted during the California Gold Rush, which began in 1848 and ended in 1855 (Maxwell-Long, 2014). With a peak in immigration due to the Rush causing 300,000 people to travel to California from all over the world, modular was the solution. As with the Australian examples of modular construction, components could be compiled or completed or assembled easily upon their arrival in America. Next, modular construction appeared in Crimea, a peninsula on the northern coast of the Black Sea in Eastern Europe (Trifonov et al., 2020a), 2020b). Following Florence Nightingale's infamous letter to The Times in 1854, Isambard Kingdom Brunel was commissioned to build a modular hospital in Scutari. Design and creation only took five months, demonstrating the speed of building modular buildings.

World War II had a significant impact on all aspects of life. During and after the war, the building of houses stopped/ceased, despite many houses becoming damaged or destroyed by bombs (Bevan, 2016). This created a housing crisis which needed a quick and efficient solution. Years later, there was the formation of the first modular housing scheme that can be traced back to Buckminster Fuller, whose flexible housing experiment of the 1920s and 1930s, the Dymaxion House, came with innovations such as notably advanced prefabricated bathroom modules (Schneiderman, 2010). These modules were shipped to the US military during World War II. However, the full extent of Fuller's modular ideas was never completely realised owing to a lack of funding in his Dymaxion company, and the engineer-architect soon moved on to other projects.

The first successful construction of a fully modular home system did not materialise until 1933, with the Winslow Ames House by Robert W. McLaughlin and his company, American House, International (Henning & Witkowski, 2013). This innovative house was built with the help of a new exterior finishing material called Cemesto, a panel board made partly of sugarcane, patented by the John Pierce Foundation. The Winslow Ames House consisted of several room modules serviced by a 'service core' to which all bathroom, kitchen, plumbing, and heating systems were attached (Cramer et al., 2019). In 1942, the US Government used a similar Cemesto prefabrication system to build the top-secret town of Oak Ridge, Tennessee, home of the Manhattan Project, virtually overnight. The government hired the firm Skidmore,

Owings, & Merrill to come up with a scheme called "Flexible Space". These were fully modular homes flexible enough for the variety of families soon to inhabit this clandestine new place. The houses came in discrete sections, cast in cement, and assembled on site in slightly varying configurations.

After the war, the concept of modularity spread to the burgeoning suburbia, albeit in a mellowed-out way, in the form of plug-and-play panel systems of the 1940s and 50s Lustron and Gunnison prefabricated houses (Nordyke, 2018). These famous houses comprised steel frames with customisable features such as various kitchen and bathroom? Systems, windows and doors, and add-ons such as porches, porticos, and garages. While only somewhat modular in design, the houses helped introduce the concepts of modularity to the general public. The modular idea captivated many throughout the 1950s, notably the architect and designer George Nelson, most famous for his mid-century modern furniture (Dodd, 2009). Nelson led the transition to new emerging concepts of modularity, expressing modular construction on the exterior (unlike his predecessors, who sought to obscure the modular nature of their designs). His ongoing Experimental House concept combined new materials, such as affordable plastics, with space-age futurism, for which Nelson was most famous.

In the 1960s, the futuristic aesthetic language pioneered by Nelson and the Eameses and popularised by television programmes such as Star Trek and current events, including the Space Race and the Atomic Age, became infused in the modular housing schemes of the 1960s (Petroli, 2021). These schemes, which focused on installing individual housing units on a mass structural frame, were forward-thinking solutions to the upcoming overpopulation crisis. The modular era of the early 60s opened with fanciful leisure housing by the French architect Guy Dessauges, whose elegant capsule houses on pylons offered panoramic views of their scenic locations (Xhexhi & Shtepani, 2023).

The concept of modularity reached an unattainable level of complexity in 1964 with "Plug-In City", a mega-scale modular concept by the British experimental architecture collective Archigram (Kiib, 2011). Though it was never built, Plug-In-City set the stage for the idea of an interchangeable city for the somewhat dystopian future predicted in the Cold War atmosphere of the 1960s. Perhaps the biggest accomplishment of this heyday of modular experimentation was the prefabricated modular megastructure by Moshe Safdie, Habitat 67, built for the 1967 Expo in Montreal (Bischeri & Micheli, 2016). Habitat 67, a graduate project by Moshe Safdie, who studied at McGill University, consists of up to 12 stories of 354 identical prefab concrete apartments arranged in various combinations. The project focused on integrating light, fresh air, and open space in the context of a dense urban community prototype. The apartments still stand and sell for hundreds of thousands of dollars (a rarity for experimental housing schemes).

On the other side of the world, Japan, confronted by a post-war population and manufacturing boom, developed the architectural style of Metabolism, which focused on flexibility, modularity, and the Archigram-like concept of interchangeable units (Cho & Shin, 2014). Though the Metabolist group was only active until the 1973 Energy Crisis, they greatly affected American architects and thinkers who visited for the 1970 Expo. Similar to Archigram, the Metabolist's concepts were primarily hypothetical, never reaching past the model stage. However, they managed to build a few buildings. The most notorious and exciting of these structures was the 1972 Nagakin Capsule Tower by Kisha Kurokawa. The Nagakin Capsule Tower consists of 140 self-contained prefabricated capsules, complete with bathrooms, cabinetry, and a built-in HiFi set (Lin, 2011). The tiny capsules, designed to be removable and replaceable, only measure 7.5 × 6.9 × 12 feet, a predecessor to today's increasingly popular micro-apartments. Despite its striking appearance, the tower has been in danger of demolition for over ten years; for now, preservation efforts have forestalled the potential loss of such an important building.

The American architect Paul Rudolph, most famous for his Brutalist government buildings, was inspired by modular projects seen in World Expositions and those by his European contemporaries (Pieris, 2011). Rudolph devised a mass-housing project in New Haven, Connecticut, near Yale, where Rudolph taught architecture. Rudolph's concept took existing housing infrastructure developments, namely the single-wide trailer, and applied it in a Habitat 67-esque fashion. Rudolph's project, Oriental Masonic Gardens, was assembled from 148 prefabricated units, grouped in clusters of four around a central utility 'core'. The living spaces were based on stacked units—the ground floor unit consisted of living spaces, with bedrooms located within the second-story unit. A third-story unit could be added as an additional living space. The project was ultimately unsuccessful. Cervero Sánchez (2022) reported that Rudolph, in a later interview, suggested that "…people hated it". First, it leaked, which is an excellent reason to hate something, but it was much more complicated than that. Psychologically, the good folk who inhabited these dwellings thought they were beneath them. In other words, the deviation of the dwelling was not something to their liking. Unsuccessful developments such as Rudolph's pushed public taste away from the mass modular concept as the 70s ended. The resolution of the 1970s energy crisis, combined with the new era of Neoliberal economic theory and prosperity created by the technological advantages of the Green Revolution, created a boom in consumerism manifested as housebuilding during the Reagan era (Wanner, 2015). Americans were all too happy to put the grim weight of social programmes and societal turmoil (including failed social housing) of the past two decades behind them and usher in a period of increased private spending and private ownership.

While modular apartment concepts retained favourability in the densely populated cities of Europe during this period, the modular housing concepts of the 1980s in America returned to the thinking of the 1940s and 50s, focusing on the individual home rather than the collective housing prototype (O'Neill & Organ, 2016). The 1980s saw modular housing technology applied to the construction of suburban homes, with companies such as Unity Homes devising modular building methods as an alternative to the often-stigmatized manufactured homes of the time. These new modular houses were constructed from parts in a factory, where a steel frame was used as the chassis on which the house was transported to the construction site (Chen et al., 2021a, 2021b))The home was picked up with a crane and assembled on a prepared foundation. The frame returned to the factory to construct the next "off frame" modular. These modular methods, whose origins are directly linked to the early cement houses of Oak Ridge, Tennessee, remain popular in-home construction today.

Lu (2019) suggested that the significant technological advances during the 1990s and 2000s inspired a new generation of architects to return to experimental modular housing. Improvements in computer rendering led to innovative manufacturing concepts such as the Klip House, whose computer-designed parts could be snapped together like LEGOs. The flexible nature of Klip House aimed to appeal to those seeking a single-family detached home that could adapt to their changing needs (Organo et al., (2013). With this system, the need for more space could be solved by simply clipping in another module rather than moving out or building an entirely new home (therefore mitigating the relentless urban sprawl of the last four decades).

The most recent developments in modular housing have sought to solve an entirely new problem: urban housing affordability (Fields & Uffer, 2014). As increased wealth moves into cities and speculation on urban property rise, displacement and gentrification have created an urban housing affordability crisis spanning cities from coast to coast. Container housing, a creative reuse approach in which excess or discarded shipping containers are transformed into tiny homes, hit the mainstream consciousness in 2000 (Islam et al., 2016) when the firm Urban Space Management completed the Container City I project in the Trinity Buoy Wharf area of London. The project is similar to Rudolph's Oriental Masonic Gardens in that several discrete units are assembled into a complex of private homes. The firm has gone on completing 16 additional projects.

Though popular in Europe, complicated zoning laws in the US have prevented the proliferation of shipping container developments despite claims that they could be an effective affordable housing solution (Ford & Gomez-Lanier, 2017). The micro-apartment, a housing solution with origins in extremely high-density cities such as Hong Kong, has spread to cities in the US, with a notable project in New York and Seattle. Projects such as My Micro New York consist of prefabricated 313 square-foot modules assembled on a steel frame on-site. An unexpected side effect of this problem is the displacement of young people and creative types, who have become a new sect of marginalised home buyers, priced out of the places where their careers can thrive. New housing types have arisen for those desperately seeking low rents: container houses and modular micro-apartments.

Lau and Wei (2018) revealed that the micro-apartment, a housing solution with origins in extremely high-density cities such as Hong Kong, has spread to cities in the US, with notable projects in New York and Seattle. Projects such as "My Micro New York" consist of prefabricated 313 square-foot modules assembled on-site on a steel frame. The micro-apartment, with its origins in projects such as Nagakin Capsule Tower, aims to provide safe living spaces with lower rents, a response to situations where bedrooms are divided ad hoc into separate dwelling spaces under the nose of the law in order to allow more inhabitants to save money on rent. Time will tell whether or not these schemes are ultimately successful, as is the case with the vast majority of mass-modular housing concepts. Mass modular housing has had a few successes, such as Habitat 67; however, most projects never make it past the planning stage or are resounding failures (Jagarajan et al., 2017). However, the plug-in-play flexibility of modular construction has always inspired forward-thinking architects. The harsh reality of our times is that the fossil-fuel-centric world of sprawl is short-lived and unsustainable. With the proliferation of new forms of housing and non-nuclear living situations, combined with our highly technological age, the conditions are ripe for a golden age of affordable modular housing, as long as zoning boards, cities, and individuals keep an open mind.

Though popular in Europe, complicated zoning laws in the US have prevented the proliferation of shipping container developments despite claims that they could be an effective, affordable housing solution (Ford & Gomez-Lanier, 2017). Factories began producing components required to create temporary housing following the war. These homes were erected throughout the UK, and many of those houses still stand today despite being temporary measures. Helping to restore the economy and provide a solution to the ongoing housing crisis, these homes were easy and cheap to build. According to the study by Ferdous et al. (2019), there is anticipated growth in the modular construction industry in the next few years. The market is projected to increase from $92.18 billion in 2018 to $130 billion in 2030. This is due, in part, to the increase in demand for rapid and affordable construction in several major industries, such as hospitals, schools, and hospitality. For example, in the UK in 2017, modular methods in building practices increased by 6%. This rapid growth is likely due to modular construction not requiring as much onsite space; this is a fundamental necessity in countries with limited space. Other cities worldwide are starting to do the same as cities become denser and more populated. Conklin Hall at Dutchess Community College is a prime example of how modular construction methods can build a large project on budget, on time, and in a limited space (Ramaji et al., 2017). Modular construction may date back two centuries; however, its processes and applications always look to tomorrow.

1.5 Objectives and Contribution of the Book

To create a modularity methodology model for the construction industry, this research study evaluated the various factors that affect how modularity approaches are incorporated into construction processes. The specific research objectives set for this study are as follows:

- To determine the factors that influence the development of a modularity methodology model for the construction industry.
- To identify the theories and models dwelling on/relating to modularity methodology for the construction industry;
- To evaluate the applications of modularity methodology for different types of buildings.
- To develop a modularity methodology model for the construction industry.
- To validate the developed modularity methodology model for the construction industry.

The achievement of the above objectives is explained in the succeeding chapter. However, it is essential that the reader is made aware at this stage as to how the above objectives have been achieved. By completing a review of the relevant literature, research goals RO1 and RO2 were met. It is essentially a theoretical knowledge of factors affecting modularity methodology. The interview was used to accomplish the RO3 goal. By using the data from the qualitative interview and the conclusions from a thorough literature analysis, research objective RO4 was successfully completed. The reasons why an interview was chosen over alternative approaches are also outlined in the parts that follow the detailed explanation of the method. Using a field questionnaire survey and the structural equation modelling (SEM) software known as EQations Software (EQS), the end aim RO5 was accomplished.

1.6 Structure of the Book

This research study was structured into six (6) separate chapters, which are titled as follows:

Chapter 1—Introduction

This chapter presents the study's background and the main research problem. The chapter also presents a general description of the study, stating the aim and objectives of the study. In addition, a description of the methods that were used to conduct this study, including the results that were obtained and ethical considerations, is presented in this chapter.

Chapter 2—Overview of Modular Construction

This chapter reviews the literature on modular construction. It gives an overview of modular construction, outlining the forms of modular construction, the history and development of modular construction, requirements for its use, and projecting the limitations to modular construction adoption.

Chapter 3—Theoretical and Conceptual Perspective of Modularity Methodology

This chapter presents a detailed review of the theories and models of modular construction that have been propounded over time. Based on their evolution, a chronological and sequential view of these theories and models is showcased. The theoretical underpinnings of the study are also outlined in this chapter.

Chapter 4—Gaps in Modularity Methodology Research

This chapter explores the identified literature gaps that bother modularity methodology. This was premised on unravelling the gaps peculiar to modularity methodology for the construction industry in developing countries. The identified gaps are reassembly, maintenance and cultural perception. These constructs were deemed crucial to the construction industry's modularity methodology practice.

Chapter 5—Global Perspective of Modularity Methodology in the Construction Industry

This chapter delves into a global perspective on modular construction. The economic theories of modular construction are examined in this chapter. Other discussions include the principles of modularity methodology, the economic analysis of modular construction and matching up modularity methodology applications in the built environment.

Chapter 6—Conceptualising Modularity Methodology Model Adoption Framework

This chapter focuses on the conceptualised model for adopting modularity methodology for the construction industry. This is attributed to a multidimensional framework comprising nine constructs: supply chain, productivity, digitalisation, legal and regulatory considerations, sustainability, procurement methods, reassembly, maintenance and cultural perception. These variables are products of a well-reviewed literature review, and the interviews conducted for the research.

Chapter 7—Validation of the Conceptual Framework

The chapter presents the analysis results based on data elicited from the questionnaire survey respondents. The comprehensive analysis based on both descriptive and inferential statistics outlined the constructs and variables required for the adoption of modular construction for the construction industry.

Chapter 8—Expert Opinions and Industry Insights Results

Chapter 8 presents the outcomes of the interviews conducted in the research. The analysis was carried out for the individual questions posed to the panel of experts. The questions dwelt on the significance of the main and sub-attributes of modularity methodology for the construction industry.

Chapter 9—Conclusion and Recommendations

This chapter maps out the conclusions drawn from the study resulting from the findings obtained from the analyses conducted. Also, recommendations are made based on the conclusions reached for the study. The contribution made by the research to the body of knowledge is presented, while areas for further studies are also recommended based on the current study's limitations.

1.7 Summary

Chapter 1 introduced the subject of the research study and gave insight into its structure, background, and significance. It provided information on how the research report is presented. The next chapter investigates the overview of modular construction.

References

Adekunle, P., Aigabvboa, C., Otasowie, K., & Akinradewo, O. (2024a). Matching-up modularity methodology application within the built environment: A bibliometric review. *Journal of Asian Architecture and Building Engineering*, 1–16. https://doi.org/10.1080/13467581.2024.2368653.

Adekunle, P., Aigbavboa, C., Otasowie, O., & Idowu, A. (2024b). Economic factors influencing the adoption of modularity methodology in the built environment. In *Transforming construction with off-site methods and technologies*.

Adekunle, P., Aigbavboa, C., Otasowie, O., Akinradewo, O., & Pheza, N. (2024c). Benefits of integrating advanced composite materials into modular construction for enhanced structural performance. In *Transforming construction with off-site methods and technologies*.

Adekunle, P., Aigabvboa, C., Thwala, W., Akinradewo, O., & Oke, A. (2022). Challenges confronting construction information management. *Frontiers in Built Environment, 8*, 1075674. https://doi.org/10.3389/fbuil.2022.1075674

Agha, A., Shibani, A., Hassan, D., & Zalans, B. (2021). Modular construction in the United Kingdom housing sector: Barriers and implications. *Journal of Architectural Engineering Technology, 10*(2).

Akinradewo, O., Aigbavboa, C., Aghimien, D., Oke, A., & Ogunbayo, B. (2023). Modular method of construction in developing countries: The underlying challenges. *International Journal of Construction Management*, 23(8), 1344–1354. https://doi.org/10.1080/15623599.2021.1970300

Amoah, C., & Simpeh, F. (2021). Implementation challenges of COVID-19 safety measures at construction sites in South Africa. *Journal of Facilities Management, 19*(1), 111–128. https://doi.org/10.1108/JFM-08-2020-0061

Armstrong, R. (2023). Towards the microbial home: An overview of developments in next-generation sustainable architecture. *Microbial Biotechnology, 16*(6), 1112–1130. https://doi.org/10.1111/1751-7915.14256

Baghchesaraei, O. R., Lavasani, H. H., & Baghchesaraei, A. (2016). Behavior of prefabricated structures in developed and developing countries. *Bulletin De La Société des Sciences De Liège, 85*, 1229–1234.

Bendi, D., Rana, M. Q., Arif, M., Goulding, J. S., & Kaushik, A. K. (2021). Understanding off-site readiness in Indian construction organisations. *Construction Innovation, 21*(1), 105–122. https://doi.org/10.1108/CI-02-2020-0016

Bertram, N., Fuchs, S., Mischke, J., Palter, R., Strube, G., & Woetzel, J. (2019). *Modular construction: From projects to products*. McKinsey & Company. https://www.mckinsey.com/industries/capital-projects-and-infrastructure/our-insights/modular-construction-from-projects-to-products

Bevan, R. (2016). *The destruction of memory: Architecture at war* (2nd ed.). Reaktion Books.

Bischeri, C., & Micheli, S. (2016). New directions for the dense city: Moshe Safdie in Singapore. *Joelho, 7*, 117–131.

Boyd, N., Khalfan, M. M. A., & Maqsood, T. (2013). Off-site construction of apartment buildings. *Journal of Architectural Engineering, 19*(1), 51–57. https://doi.org/10.1061/(asce)ae.1943-5568.0000091

Campagnolo, D., & Camuffo, A. (2010). The concept of modularity in management studies: A literature review. *International Journal of Management Reviews, 12*(3), 259–283. https://doi.org/10.1111/j.1468-2370.2009.00260.x

Cervero Sánchez, N. (2022). Paul Rudolph's construction: Material and space. In C. Bartolomei, A. Ippolito, & S. H. T. Vizioli (Eds.), *Digital modernism heritage lexicon* (pp. xxx–xxx). Springer. https://doi.org/10.1007/978-3-030-76239-1_41

Chau, C. K., Leung, T. M., & Ng, W. Y. (2015). A review on life cycle assessment, life cycle energy assessment and life cycle carbon emissions assessment on buildings. *Applied Energy, 143*(1), 395–413. https://doi.org/10.1016/j.apenergy.2015.01.023

Chen, Z., Song, P., & Wang, B. (2021a). Carbon emissions trading scheme, energy efficiency and rebound effect: Evidence from China's provincial data. *Energy Policy, 157,* 112507. https://doi.org/10.1016/j.enpol.2021.112507

Chen, Z., Song, P., & Wang, B. (2021b). Carbon emissions trading scheme, energy efficiency and rebound effect—Evidence from China's provincial data. *Energy Policy, 157*, 112507.

Chen, Y., Okudan, G. E., & Riley, D. R. (2010). Sustainable performance criteria for construction method selection in concrete buildings. *Automation in Construction, 19*(2), 235–244. https://doi.org/10.1016/j.autcon.2009.10.004

Cho, H., & Shin, C. (2014). Metabolism and cold war architecture. *The Journal of Architecture, 19*(5), 623–644. https://doi.org/10.1080/13602365.2014.965186

Cramer, N. B., Cellucci, D. W., Formoso, O. B., Gregg, C. E., Jenett, B. E., Kim, J. H., Lendraitis, M., Swei, S. S., Trinh, G. T., Trinh, K. V., & Cheung, K. C. (2019). Elastic shape morphing of ultralight structures by programmable assembly. *Smart Materials and Structures, 28*(5), Article 055006. https://doi.org/10.1088/1361-665X/ab0ea2

Cusumano, M. A., Kahl, S. J., & Suarez, F. F. (2015). The Effect of firm compensation structures on the mobility and entrepreneurship of extreme performers. *Strategic Management Journal, 36*, 559–575. https://doi.org/10.1002/smj

Davila Delgado, J. M., Oyedele, L., Ajayi, A., Akanbi, L., Akinade, O., Bilal, M., et al. (2019). Robotics and automated systems in construction: Understanding industry-specific challenges for adoption. *Journal of Building Engineering, 26*, 100868. https://doi.org/10.1016/j.jobe.2019.100868

Dodd, S. T. (2009). Merchandising the postwar model house at the Parade of Homes (Unpublished M.A. thesis). University of Texas at Austin.

Ferdous, W., Bai, Y., Ngo, T. D., Manalo, A., & Mendis, P. (2019). New advancements, challenges and opportunities of multi-storey modular buildings—A state-of-the-art review. *Engineering Structures, 183*, 883–893. https://doi.org/10.1016/j.engstruct.2019.01.061

Fields, D., & Uffer, S. (2014). The financialisation of rental housing: A comparative analysis of New York City and Berlin. *Urban Studies, 53*(7), 1486–1502. https://doi.org/10.1177/0042098014543704

Ford, J., & Gomez-Lanier, L. (2017). Are tiny homes here to stay? A review of literature on the tiny house movement. *Family and Consumer Sciences Research Journal, 45*(4), 394–405. https://doi.org/10.1111/fcsr.12205

Gal, U., Lyytinen, K., & Yoo, Y. (2008). The dynamics of IT boundary objects, information infrastructures, and organisational identities: The introduction of 3D modelling technologies into the architecture, engineering, and construction industry. *European Journal of Information Systems, 17*(3), 290–304. https://doi.org/10.1057/ejis.2008.13

Gálvez-Martos, J., Styles, D., Schoenberger, H., & Zeschmar-Lahl, B. (2018). Construction and demolition waste best management practice in Europe. *Resources, Conservation and Recycling, 136*, 166–178. https://doi.org/10.1016/j.resconrec.2018.04.016

George, A. S., & George, A. S. H. (2020). Industrial revolution 5.0: The transformation of the modern manufacturing process to enable man and machine to work hand in hand. *Journal of Seybold Report, 15*(9), 214–234. https://doi.org/10.5281/zenodo.6548092

Gershenson, J. K., Prasad, G. J., & Zhang, Y. (2004). Product modularity: Measures and design methods. *Journal of Engineering Design, 15*(1), 33–51. https://doi.org/10.1080/0954482032000101731

Ghaffar, S. H., Corker, J., & Fan, M. (2018). Additive manufacturing technology and its implementation in construction as an eco-innovative solution. *Automation in Construction, 93*, 1–11. https://doi.org/10.1016/j.autcon.2018.05.005

Giorgio, B., Blanchet, P., & Barlet, A. (2022). Social representations of mass timber and prefabricated light-frame wood construction for multi-story housing: The vision of users in Quebec. *Buildings, 12(*12). https://doi.org/10.3390/buildings12122073

Groumpos, P. P. (2020). A critical historical and scientific overview of all industrial revolutions. *IFACPapersOnLine, 54*(13), 464–471. https://doi.org/10.1016/j.ifacol.2021.10.492

Henning, R. A., & Witkowski, T. H. (2013). The advertising of E. Remington & Sons: The creation of an iconic brand, 1854–1888. *Journal of Historical Research in Marketing, 5*(4), 418–448. https://doi.org/10.1108/JHRM-11-2012-0028

Ilhan, B., & Yaman, H. (2016). Green building assessment tool (GBAT) for integrated BIM-based design decisions. *Automation in Construction, 70*, 26–37. https://doi.org/10.1016/j.autcon.2016.05.001

Islam, H., Zhang, G., Setunge, S., & Bhuiyan, M. A. (2016). Life cycle assessment of shipping container home: A sustainable construction. *Energy and Buildings, 128*, 673–685. https://doi.org/10.1016/j.enbuild.2016.07.002

Jagarajan, R., Abdullah Mohd Asmoni, M. N., Mohammed, A. H., Jaafar, M. N., Lee Yim Mei, J., & Baba, M. (2017). Green retrofitting—A review of current status, implementations and challenges. *Renewable and Sustainable Energy Reviews, 67*, 1360–1368. https://doi.org/10.1016/j.rser.2016.09.091

Jiang, Y., Li, M., Guo, D., Wu, W., Zhong, R. Y., & Huang, G. Q. (2022). Digital twin-enabled smart modular integrated construction system for on-site assembly. *Computers in Industry, 136*, 103594. https://doi.org/10.1016/j.compind.2021.103594

Kamali, M., & Hewage, K. (2016). Life cycle performance of modular buildings: A critical review. *Renewable and Sustainable Energy Reviews, 62*, 1171–1183. https://doi.org/10.1016/j.rser.2016.05.031

Khadodra, U., Rahaman, M. H., & Jamil, M. (2020). Design of renewable energy system for a house in St. John's, Canada. *European Journal of Electrical Engineering and Computer Science, 4*(3). https://doi.org/10.24018/ejece.2020.4.3.213

Kiib, H. (2011). Deturned city design as tool for aesthetic urban living (Master's thesis). Aalborg University. http://vbn.aau.dk/ws/files/63285525/Deturned_City_Design_final_ver_2.pdf

Kineber, A. F., Oke, A. E., Hamed, M. M., Rached, E. F., Elmansoury, A., & Alyanbaawi, A. (2023). A partial least squares structural equation modeling of robotics implementation for sustainable building projects: A case in Nigeria. *Sustainability* (Switzerland), *15*(1). https://doi.org/10.3390/su15010604

Kolbeck, L., Kovaleva, D., Manny, A., Stieler, D., Rettinger, M., Renz, R., et al. (2023). Modularisation strategies for individualised precast construction—Conceptual fundamentals and research directions. *Designs, 7*(6), 143. https://doi.org/10.3390/designs7060143

Krishna, M. R. V. (2021). Role of Industry 4.0 in reshaping the future of the industry. *Journal of development research, 14*(3), 16.

Lau, M. H. M., & Wei, X. (2018). Housing size and housing market dynamics: The case of micro-flats in Hong Kong. *Land Use Policy, 78*, 278–286. https://doi.org/10.1016/j.landusepol.2018.06.039

Lawson, R. M., Ogden, R. G., & Bergin, R. (2012). Application of modular construction in high-rise buildings. *Journal of Architectural Engineering, 18*(2), 148–154. https://doi.org/10.1061/(ASCE)AE.1943-5568.0000057

Lin, Z. (2011). Nakagin capsule tower: Revisiting the future of the recent past. *Journal of Architectural Education, 65*(1), 13–32. https://doi.org/10.1111/j.1531-314x.2011.01158.x

Lu, Y. (2019). Artificial intelligence: A survey on evolution, models, applications and future trends. *Journal of Management Analytics, 6*(1), 1–29. https://doi.org/10.1080/23270012.2019.1570365

MacDuffie, J. P. (2013). Modularity-as-property, modularization-as-process, and 'modularity'-as-frame: Lessons from product architecture initiatives in the global automotive industry. *Global Strategy Journal, 3*(1), 8–40. https://doi.org/10.1111/j.2042-5805.2012.01048.x

Maxwell-Long, T. (2014). *Daily life during the California Gold Rush* (Illustrated ed.). ABC-CLIO.

McLeod, L., Doolin, B., & MacDonell, S. G. (2012). A perspective-based understanding of project success. *Project Management Journal, 43*(5), 68–86. https://doi.org/10.1002/pmj.21290

Murali, K., & Sambath, K. (2020). Sustainable performance criteria for prefabrication construction system. *International Journal of Scientific and Research Publications, 10*, 455–458.

Navaratnam, S., Ngo, T., Gunawardena, T., & Henderson, D. (2019). Performance review of prefabricated building systems and future research in australia. *Buildings, 9*(2), 38. https://doi.org/10.3390/buildings9020038

Ndekugri, I., Ankrah, N. A., & Adaku, E. (2022). The design coordination role at the pre-construction stage of construction projects. *Building Research & Information, 50*(4), 452–466. https://doi.org/10.1080/09613218.2021.1971061

Ngowi, A. B., Pienaar, E., Talukhaba, A., & Mbachu, J. (2005). The globalisation of the construction industry—A review. *Building and Environment, 40*(1), 135–141. https://doi.org/10.1016/j.buildenv.2004.05.008

Nordyke, M. G. (2018). Restyling the postwar prefab. *Buildings & Landscapes: Journal of the Vernacular Architecture Forum, 25*(2), 66–94.

Onat, N. C., Kucukvar, M., & Tatari, O. (2014). Scope-based carbon footprint analysis of U.S. residential and commercial buildings : An input–output hybrid life cycle assessment approach. *Building and Environment, 72*, 53–62.

O'Neill, D., & Organ, S. (2016). A literature review of the evolution of British prefabricated low-rise housing. *Structural Survey, 34*(2), 140–157. https://doi.org/10.1108/SS-08-2015-0037

Organo, V., Head, L., & Waitt, G. (2013). ¿Quién hace el trabajo en los hogares sustentables? Un análisis de tiempo y género en la Nueva Gales del Sur, Australia. *Gender, Place & Culture, 20*(5), 559–577. https://doi.org/10.1080/0966369X.2012.716401

Otasowie, O.K., Aigbavboa, C.O., Oke, A.E. & Adekunle, P. (2024), Mapping out focus for circular economy business models (CEBMs) research in construction sector studies–a bibliometric approach, *Journal of Engineering, Design and Technology,* Vol. ahead-of-print No. ahead-of-print. https://doi.org/10.1108/JEDT-10-2023-0444

Otasowie, O., Aigbavboa, C., Oke, A., & Adekunle, P. (2025). Redefining key activities in construction organisations: towards the adoption of circular economy business model. *International Journal of Construction Management,* 1–14. https://doi.org/10.1080/15623599.2025.2554940

Pedercini, M., Zuellich, G., Dianati, K., & Arquitt, S. (2018). Toward achieving sustainable development goals in Ivory Coast: Simulating pathways to sustainable development. *Sustainable Development, 26*(6), 588–595. https://doi.org/10.1002/sd.1721

Pervez, H., Ali, Y., & Petrillo, A. (2021). A quantitative assessment of greenhouse gas (GHG) emissions from conventional and modular construction: A case of developing country. *Journal of Cleaner Production, 294*, 126210. https://doi.org/10.1016/j.jclepro.2021.126210

Petroli, M. A. (2021). *Toward a modern civic monumentality: Arches, vaults, and domes in postwar American architecture.* Illinois Institute of Technology. https://doi.org/10.1080/09638288.2019.1595750

Pieris, A. (2011). "Tropical" cosmopolitanism? The untoward legacy of the American style in postindependence Ceylon/Sri Lanka. *Singapore Journal of Tropical Geography, 32*(3), 332–349. https://doi.org/10.1111/j.1467-9493.2011.00436.x

Pizarro, P. N., & Massone, L. M. (2021). Structural design of reinforced concrete buildings based on deep neural networks. *Engineering structures, 241*, 112377.

Ramaji, I. J., Memari, A. M., & Messner, J. I. (2017). Product-oriented information delivery framework for multistory modular building projects. *Journal of Computing in Civil Engineering, 31*(4). https://doi.org/10.1061/(asce)cp.1943-5487.0000649

Regona, M., Yigitcanlar, T., Xia, B., & Li, R. Y. M. (2022). Opportunities and adoption challenges of AI in the construction industry: A PRISMA review. *Journal of Open Innovation: Technology, Market, and Complexity, 8*(1), 45. https://doi.org/10.3390/joitmc8010045

Rosarius, A., & García de Soto, B. (2021). On-site factories to support lean principles and industrialized construction. *Organization, Technology & Management in Construction: An International Journal, 13*(1), 2353–2366.

Russell, A. L. (2012). Modularity: An interdisciplinary history of an ordering concept. *Information & Culture* 47(3), 257–287. https://doi.org/10.7560/IC47301

Russell-Smith, S. V., Lepech, M. D., Fruchter, R., & Littman, A. (2015). Impact of progressive sustainable target value assessment on building design decisions. *Building and Environment, 85*, 52–60. https://doi.org/10.1016/j.buildenv.2014.11.011

Rymarczyk, J. (2020). Technologies, opportunities and challenges of the industrial revolution 4.0: Theoretical considerations. *Entrepreneurial business and economics review, 8*(1), 185–198.

Saad, A. M., Dulaimi, M., Zulu, S. L., & Gorse, C. (2023). Are the characteristics of public clients influencing their uptake of the modern methods of construction? A thematic analysis. *Smart and Sustainable Built Environment*. https://doi.org/10.1108/SASBE-11-2022-0255

Schneiderman, D. (2010). The prefabricated kitchen: Substance and surface. *Home Cultures, 7*(3), 243–262. https://doi.org/10.2752/175174210X12785760502135

Şeker, F., & Aydinli, B. (2023). A trial patch to sustainable development. *Science & Education, 32*, 421–446. https://doi.org/10.1007/s11191-021-00315-x

Shin, J., Moon, S., Cho, B. ho, Hwang, S., & Choi, B. (2022). Extended technology acceptance model to explain the mechanism of modular construction adoption. *Journal of Cleaner Production, 342*, 130963. https://doi.org/10.1016/j.jclepro.2022.130963

Sievers, S., Seifert, T., Schembecker, G., & Bramsiepe, C. (2016). Methodology for evaluating modular production concepts. *Chemical Engineering Science, 155*, 153–166. https://doi.org/10.1016/j.ces.2016.08.006

Tam, V. W., Kotrayothar, D., & Xiao, J. (2015). Long-term deformation behaviour of recycled aggregate concrete. *Construction and Building Materials, 100*, 262–272. https://doi.org/10.1016/j.conbuildmat.2015.10.013

Thai, H., Ngo, T., & Uy, B. (2020). A review on modular construction for high-rise buildings. *Structures, 28*, 1265–1290.

Trifonov, V. G., Sokolov, S. Y., Sokolov, S. A., & Hessami, K. (2020a). Mesozoic–Cenozoic structure of the Black Sea–Caucasus–Caspian region and its relationships with the upper mantle structure. *Geotectonics, 54*(3), 331–355. https://doi.org/10.1134/S0016852120030103

Trifonov, T., Pashova, S., & Andreev, S. (2020b). Modular construction in Bulgaria—A case study. *IOP Conference Series: Materials Science and Engineering, 869*, 052021. https://doi.org/10.1088/1757-899X/869/5/052021

Tsz Wai, C., Wai Yi, P., Ibrahim Olanrewaju, O., Abdelmageed, S., Hussein, M., Tariq, S., & Zayed, T. (2023a). A critical analysis of benefits and challenges of implementing modular integrated construction. *International Journal of Construction Management, 23*(4), 656–668. https://doi.org/10.1080/15623599.2021.1907525

Tuvayanond, W., Prasittisopin, L., Tuvayanond, W., & Prasittisopin, L. (2023). Design for manufacture and assembly of digital fabrication and additive manufacturing in construction: A review. *Buildings, 13*(2). https://doi.org/10.3390/buildings13020429

Veilleux, G., Potisat, T., Pezim, D., Ribback, C., Ahmed, A., & Papenheim, J. (2020). Energy for sustainable development techno-economic analysis of microgrid projects for rural electri fi cation: A systematic approach to the redesign of Koh Jik off- grid case study. *Energy for Sustainable Development, 54*, 10–14. https://doi.org/10.1016/j.esd.2019.09.007

Wanner, T. (2015). The new 'passive revolution' of the green economy and growth discourse: Maintaining the 'sustainable development' of neoliberal capitalism. *New Political Economy, 20*(1), 21–41. https://doi.org/10.1080/13563467.2013.866082

Wrigley, P., Wood, P., O'Neill, S., Hall, R., & Robertson, D. (2021). Off-site modular construction and design in nuclear power: A systematic literature review. *Progress in Nuclear Energy, 134*, 103664. https://doi.org/10.1016/j.pnucene.2021.103664

Wuni, I. Y., & Shen, G. Q. (2020a). Barriers to the adoption of modular integrated construction: Systematic review and meta-analysis, integrated conceptual framework, and strategies. *Journal of Cleaner Production, 249*, 119347. https://doi.org/10.1016/j.jclepro.2019.119347

Wuyts, W., Miatto, A., Sedlitzky, R., & Tanikawa, H. (2019). Extending or ending the life of residential buildings in Japan : A social circular economy approach to the problem of short-lived constructions. *Journal of Cleaner Production, 231*, 660–670. https://doi.org/10.1016/j.jclepro.2019.05.258

Xhexhi, K., & Shtepani, E. (2023). Administrative modular buildings/co-working spaces. A review.

Xu, Z., Zayed, T., & Niu, Y. (2020). Comparative analysis of modular construction practices in mainland China, Hong Kong and Singapore. *Journal of Cleaner Production, 245*, 118861. https://doi.org/10.1016/j.jclepro.2019.118861

Zhang, X., Skitmore, M., & Peng, Y. (2014). Exploring the challenges to industrialized residential building in China. *Habitat International, 41*, 176–184. https://doi.org/10.1016/j.habitatint.2013.08.005

Zhou, W., Whyte, J., & Sacks, R. (2012). Construction safety and digital design: A review. *Automation in Construction, 22*, 102–111. https://doi.org/10.1016/j.autcon.2011.07.005

Part II
Overview of Modular Construction

Chapter 2
Modularity Methodology

Abstract This chapter provides a comprehensive overview of modular construction, focusing specifically on its methodological aspects. It begins by examining the various forms of modular construction methodology, highlighting how these differ in their design principles, manufacturing processes, and assembly techniques. The discussion covers key approaches such as volumetric modular construction, panelized systems, and hybrid methods, each offering distinct advantages in terms of speed, quality control, and adaptability to different project requirements. The chapter then explores the essential requirements for implementing modular construction effectively. These include project requirements, building component standardisation, and construction industry standards. Attention is also given to manufacturing capabilities, construction processes, and cost and time constraints. The chapter also addresses the limitations and challenges that hinder the widespread adoption of modular construction. Key constraints include perception, stigmatisation, limited design flexibility, transportation and complexities, and difficulties in adapting standardised modules to diverse local codes and environmental conditions. Resistance to change within the industry and the need for specialised skills further compound these challenges.

2.1 Introduction

Modular construction has numerous advantages over traditional methods (Otasowie et al., 2024a, 2024b). On-site projects are often subjected to the whims of Mother Nature; rain, extreme temperatures, and more can interfere with building projects. Modular manufacturing processes do not have that problem. The processes can even continue operation during the cold winter months, often considered the off-season, so they can be ready to start assembling again as soon as the weather warms up. Modular construction also helps improve quality control (Shin & Choi, 2022). The steps are the same every time, unlike traditional methods. Much of the process could even be automated, eliminating the risk of human error. Structural building components are a small part of a growing offsite construction trend, but people will likely see more in the coming years (Lehmann, 2012). The industry is shifting toward prefabricated

P. Adekunle et al., *Modularity Methodology*,
https://doi.org/10.1007/978-3-032-12918-5_2

techniques and off-site assembly to reduce costs and shrink its carbon footprint. It might not be long before all major building projects rely on off-site component manufacturing.

2.2 Forms of Modularity Construction Methodology

Modular construction involves planning, designing, fabrication, transporting fabricated building items and assembling them on the site with great speed and a high degree of finish (Hsu et al., 2018). This degree of precision is far better than on-site construction. Modular building construction incorporates various materials, scales and systems, innovative manufacturing methods and fabrication, digital software and modern assembling techniques. The output from a modular construction has panellised and modularised components that perform as structural elements, enclosure and service elements, or interior partition systems (Isaac et al., 2016). Optimisation in off-site construction is achieved by integrating these systems and the supply chain with proper and efficient research, design, testing and prototyping.

There are different forms through which an offsite construction can be performed. Ayinla et al. (2020) classified these as volumetric construction, hybrid construction, panellised construction, and sub-assemblies and accessories systems. In volumetric construction, three-dimensional units are produced in the factory, brought to the site, and bolted together. The frames constructed can be either steel, timber or concrete. The factory possesses quality-controlled production systems as a part of third-party approval. The modules in volumetric construction can be bought in different forms, such as the bare structural shell or the one with internal and external finishes and services (Boyd et al., 2013). This construction may include timber concrete, light gauge steel frame or composite materials. Volumetric construction can be further classified into pod construction. In pod construction, factory-produced pre-engineered building components are delivered to the construction site (Williamson et al., 2019). These modules are then assembled as significant volumetric components or as substantial elements. Pods are non-structural elements that are used with a load-bearing structure. They are mainly employed in the construction market of hotels and accommodation spaces. The enclosure can be a timber, concrete, steel, or load-bearing structure. Also, in this type of construction, three-dimensional elements are employed in the superstructure of the building (Patil et al., 2021). These make ready-made rooms that can be combined to obtain complete premises. These are set together within a light steel framework. All the final building services are pre-installed at the site, which involves the final connection—for example, the construction of hotel bathrooms, accommodation blocks and kitchen units.

Panellised construction, the second form of modular construction, refers to flat panel units used as walls, floors or roof panes to create a complete structural shell (Thai et al., 2020). These panel units are built in the factory and transported to the site to be assembled as a three-dimensional structure or to fit into an already constructed structure on site. The factory-made structural floors and roofs are also called cassettes.

These cassettes can be either load-bearing or non-load-bearing structures. These structural panel units can be made from timber, insulated panels, light gauge steel, concrete, or non-structural in-fill walls (Deng et al., 2020). Any building can make use of such panel units. Some of the examples of building components constructed using panellised systems are windows, timber frames, insulation, and windows. The panellised construction can use open, closed, concrete, insulated, composite, and infill panels.

The third form of modular construction, hybrid construction, combines volumetric and panellised approaches to develop a single building (Ginigaddara et al., 2022). Hence, this construction is also called semi-volumetric. The hybrid systems are either building facilities that are entirely manufactured or prefabricated in the factory—these units have the complete final finish. The hybrid systems, called pods, are transported to the site after finishing (Ginigaddara et al., 2019). For example, highly serviced areas such as bathrooms and kitchens are initially constructed as volumetric units, whereas the rest of the dwelling is constructed with the help of panels.

Finally, the last but not the most diminutive form of modular construction, termed sub-assemblies and accessories system, is the process by which the building components are initially constructed at the offsite location before they are permanently erected at the site (Tzourmakliotou, 2021). These elements include the building components, materials, equipment and prefabricated parts. This construction involves manufacturing small components such as roof trusses, floor piping, staircases, roof trusses, precast concrete beams and columns. The sub-assembly system consists of pre-fabricated foundations, floor cassettes and roof cassettes.

2.3 Requirements of Modularity Construction Methodology

Modularity methodology used in construction is based on various requirements that influence the degree of modularity in building projects (Sharafi et al., 2018). These determinants shape the selection and application of modular construction methodologies. It delves into the interplay among project requirements, industry standards, manufacturing capabilities, construction processes, and cost and time constraints. Each determinant influences the decision-making process, ultimately shaping the modularity approach employed in a construction project (Shahpari et al., 2020). The requirements discussed in this context serve as guiding principles for stakeholders involved in the design, planning, and execution of modular construction projects. By examining these requirements, architects, engineers, contractors, and developers can make informed choices that align with project goals, optimise resource allocation, enhance efficiency, and deliver high-quality built environments.

It is important to note that the requirements are not isolated factors but interconnected elements that require a holistic approach (Baldassano & Bassett, 2016).

They influence one another and must be carefully balanced to ensure the successful implementation of modular construction. Stakeholders can successfully negotiate the challenges of modular construction and realise all its advantages by acknowledging each factor's importance and combined influence (Qi et al., 2021). An explanation of these prerequisites provides a thorough overview of the factors influencing the uptake and utilisation of modular building. Stakeholders can traverse the decision-making process and take advantage of the revolutionary potential of modularity in construction by assessing project needs, industry standards, manufacturing capabilities, building procedures, and cost and time restrictions.

2.3.1 Project Requirements

The project requirements dictate the degree of customisation and flexibility required in building design (Lee & Hyun, 2019). Projects with highly customised requirements, such as luxury residences or speciality buildings, may require lower levels of modularity than projects with more standardised requirements, such as office buildings or apartments. Project requirements are crucial in determining the most suitable modularity methodology for a construction project (Kim & Lee, 2014). The specific requirements of a project, including its scope, timeline, budget, functionality, and site constraints, can significantly influence the choice of modular construction approaches. The scale of the project, such as the number of units or the overall size of the structure, can impact the selection of modularity methodology. Volumetric modular construction might be more efficient for larger projects, where complete modules are fabricated off site and assembled on site (Tasan & Gen, 2008). On the other hand, smaller projects may benefit from panellised modular construction, where components are manufactured and assembled on site.

Furthermore, if the project has a tight schedule or requires rapid construction, modular construction can offer significant time savings compared to traditional construction methods (Taghaddos et al., 2014). The level of modularity, such as the extent to which modules or components are prefabricated off site, can be adjusted based on the project timeline to optimise speed and efficiency. Another factor to consider for project requirements is design flexibility. The design requirements of the project can influence the modularity methodology chosen. Projects with more standardised or repetitive design elements can easily leverage modular systems based on predefined module sizes and configurations (Browning & Yassine, 2010). However, projects requiring unique or custom designs might require more flexible modular approaches to accommodate specific design specifications. The characteristics and limitations of the construction site, known as site constraints, can also impact the selection of modularity methodology. Lawson and Ogden (2008) opined that modular construction can provide advantages if the site has limited access or is constrained by space restrictions, transportation challenges, or environmental concerns. Off-site fabrication and assembly can help minimise disruption at the site and address logistical challenges more effectively.

In addition, the intended functionality of the building or structure can influence the choice of modularity methodology (Moghadam et al., 2012). Certain building types, such as healthcare facilities or educational institutions, have specific requirements for functional spaces and equipment installations. Modular construction systems that integrate specialised mechanical, electrical, and plumbing (MEP) systems can be chosen to meet these functional requirements. Project budget and cost considerations also play a role in determining the modularity methodology. Modular construction can save costs through reduced labour costs, improved construction efficiency, and better control over material waste (Otasowie et al., 2024a). The choice between different modularity methodologies may depend on their cost-effectiveness for a particular project (Alvanchi et al., 2012). Project requirements are a guiding factor in selecting the most appropriate modularity methodology for a construction project. By closely analysing the project's particular needs, stakeholders can ascertain the degree of modularity, manufacturing procedures, and assembly techniques that align with the project's objectives, limitations, and intended results.

2.3.2 Building Component Standardisation

The standardisation of building components is critical in determining the degree of modularity in a construction project (Yakimchuk et al., 2023). Standardised components are designed to be interchangeable and easily assembled on-site. The degree of standardisation affects the level of modularity in the construction process. Building component standardisation is a crucial determinant of modularity methodology in construction projects (Economidou et al., 2020). Standardisation establishes uniform specifications, dimensions, and interface-building components or modules. When components are standardised, they are designed and manufactured to conform to a set of predefined specifications, allowing for interchangeability and compatibility across different projects. The first-way building component standardisation influences modularity methodology is interchangeability and compatibility (D'Agostino et al., 2021). Standardised building components enable interchangeability, meaning they can be easily swapped or replaced with similar components without requiring significant modifications. This interchangeability allows for greater flexibility in assembling and reconfiguring modules or components, promoting modularity in construction. Components with standardised dimensions and interfaces can fit together seamlessly, making assembly and disassembly processes more efficient.

The second way is through manufacturing efficiency. Standardised components simplify the manufacturing process by providing precise design specifications and reducing the need for custom fabrication (Otasowie et al., 2024b). Manufacturers can establish production lines and processes optimised for specific standardised components, resulting in improved efficiency, reduced costs, and faster production times. This efficiency is particularly beneficial in modular construction, where multiple identical or similar components are produced. Quality control and assurance is another way building component standardisation influences modularity methodology (Bastian et al., 2022). Standardised components promote quality control and assurance throughout the construction process. Quality standards can be consistently applied across different projects since the components are manufactured based on established specifications. Manufacturers can implement rigorous quality control measures, ensuring that all standardised components meet the required performance criteria (Risholt et al., 2013). This standardised approach enhances the overall quality of the modular construction system.

In addition, design consistency, procurement and supply chain management, and adaptability and scalability are also regarded as building component standardisation factors influencing modularity methodology. Standardised components contribute to design consistency across different modules or buildings within a project or even across various projects (Begic & Krstic, 2022). Modular buildings can have their structural features, aesthetic appeal, and utility preserved using uniform components. This is especially helpful for large-scale or multi-phase projects where uniformity and continuity in the design are sought. Standardised building components facilitate procurement and supply chain management (Otasowie et al., 2023). Estimating quantities, planning deliveries, and managing inventory become easier when components are standardised.

Additionally, standardisation encourages competition among suppliers, as multiple manufacturers can produce the same component based on the established specifications. This competition can reduce costs and improve the availability of standardised components in the market. Standardised building components enhance adaptability and scalability in modular construction. As standardised components can be easily integrated or replaced, modular buildings can be expanded, reconfigured, or modified as needed (Antonini et al., 2014). This adaptability and scalability are particularly valuable in projects where future modifications or additions are anticipated. Therefore, building component standardisation provides a solid foundation for modularity methodology in construction. The industry can improve productivity, quality control, design consistency, and modular construction adaptability by establishing uniform specs, measurements, and interfaces.

2.3.3 Construction Industry Standards

Industry standards for construction components and systems can influence the degree of modularity in a building project (Xu et al., 2020). Standards can guide component design, manufacturing, and assembly, facilitating using modular components in construction. The construction industry standard, which refers to commonly accepted practices, guidelines, and specifications within the construction sector, can significantly influence the selection and implementation of modularity methodology (Hong et al., 2018). Industry organisations, regulatory bodies, and professional associations develop and uphold these standards to ensure consistency, quality, and safety in construction projects. The processes through which construction industry standards act as requirements for modularity methodology include design guidelines, material specifications, building codes and regulations, quality assurance and control, safety guidelines, and documentation and project management standards.

Quale et al. (2012) opined that the construction industry standard often provides design guidelines that outline best practices and specifications for various aspects of construction. These guidelines may include recommendations for structural design, architectural considerations, mechanical, electrical, and plumbing (MEP) systems, and other relevant aspects. Adherence to these design guidelines is essential when adopting modularity methodology to ensure compatibility, safety, and compliance with industry standards. Also, construction industry standards provide specifications for various materials used in construction projects (Chen et al., 2017). These specifications include material properties, performance requirements, and testing procedures. When selecting components or modules for modular construction, adherence to industry-standard material specifications ensures the chosen materials' quality, durability, and compatibility. This promotes consistency and reliability in modular construction processes (Tavares et al., 2019). In addition, building codes and regulations are an essential part of the construction industry standard. They are established by governmental authorities to ensure the safety, health, and welfare of building occupants and the general public. When implementing modularity methodology, compliance with building codes and regulations is crucial (Liu & Qian, 2019). These codes address various aspects of construction, including structural integrity, fire safety, energy efficiency, accessibility, and more. Modular construction processes must meet these requirements to obtain necessary permits and approvals.

Furthermore, the construction industry standard encompasses quality assurance and control measures to ensure construction projects meet specified standards and requirements (Sanches et al., 2018). This includes procedures for inspections, testing, documentation, and verification. Adherence to quality assurance and control standards is vital when adopting modularity methodology to ensure the consistent production and assembly of modules or components. These standards help verify that the manufacturing processes, materials, and installation methods meet the desired quality and performance criteria (Zhang et al., 2017). Safety guidelines are also an integral part of the construction industry standard. They establish best practices for construction site safety, worker protection, and hazard mitigation. Modularity methodology need to align with these safety guidelines to ensure a safe working environment during the fabrication and assembly of modules (Hammad et al., 2019). Guidelines may address lifting and transporting modules, site logistics, handling modular components, and other safety considerations specific to modular construction. The construction industry standard must encompass documentation requirements and project management standards. These standards ensure that construction projects are adequately planned, documented, and managed (Kamali & Hewage, 2017). Proper documentation includes design drawings, specifications, contracts, change orders, and record-keeping. Adhering to project management standards helps facilitate efficient coordination, communication, and collaboration among stakeholders involved in modular construction projects. Stakeholders could ensure that modularity methodology are applied within an established framework of safety measures, quality standards, and best practices by considering and adhering to the building industry standard. This adherence helps to ensure that the modular construction process is reliable, consistent, and compliant.

2.3.4 Manufacturing Capabilities

Turner et al. (2021) referred to manufacturing capabilities as the capacity, resources, and expertise of a manufacturing facility to produce modular components or modules. These capabilities are crucial in determining the suitable modularity methodology for a construction project. The capabilities of manufacturers to produce standardised building components can impact the degree of modularity in a building project (Ogunbiyi et al., 2014). Modularity may be lower if manufacturers have limited capabilities to produce standardised components. The first factor through which manufacturing capabilities can be asked as a requirement of modularity methodology is the production capacity (Innella et al., 2019). The production capacity of a manufacturing facility is a critical consideration when determining modularity methodology. The facility's ability to produce a sufficient quantity of modular components or modules within the project timeline is essential (Baldea et al., 2017). Large-scale projects may require high production volumes, so manufacturing facilities with significant capacity are preferable. The production capacity should align with the project's requirements to ensure timely delivery and installation of modular elements.

Another factor by which manufacturing capabilities can act as a requirement for modularity methodology is the manufacturing process. Different modularity methodology may require specific manufacturing processes (Hyun et al., 2020). For example, volumetric modular construction involves fabricating complete modules off-site, requiring a facility with advanced manufacturing capabilities such as assembly lines, automation, and large-scale fabrication equipment. On the other hand, panellised modular construction involves producing individual panels or components that can be assembled on-site (Boafo et al., 2016). The manufacturing capabilities must align with the selected modularity methodology to ensure efficient and accurate production of modular elements. Design flexibility is a manufacturing capability factor required for modularity methodology as it influences modular components or modules. The facility's ability to accommodate custom designs, material variations, and aesthetic options can impact the choice of modularity methodology (Mao et al., 2013). Some manufacturing facilities may specialise in producing standardised modular components, while others can accommodate unique design specifications. The manufacturing capabilities should align with the desired design flexibility of the project.

Furthermore, manufacturing capabilities are vital in ensuring quality control and assurance throughout production (Wuni & Shen, 2020). Facilities with robust quality control measures, such as inspections, testing protocols, and certification processes, are preferable for modular construction. The manufacturing capabilities should include quality assurance processes to ensure that modular components or modules meet the required standards and specifications. The manufacturing facility's expertise in handling specific materials is also an important consideration. Modular construction can involve steel, wood, concrete, or composite (Lehmann, 2012). Manufacturing facilities with experience and expertise in working with specific materials can ensure the accurate fabrication and assembly of modular elements. The manufacturing capabilities should align with the desired materials for the project to ensure optimal performance and durability. Finally, some modularity methodology require integrating building systems within the modular components or modules, including MEP systems (Gbadamosi et al., 2019). The manufacturing capabilities should include the expertise and resources to accommodate the installation and integration of these systems. This may involve working with specialised trades or having the capability to incorporate prefabricated MEP components into the modular elements. Stakeholders can guarantee that the modularity approaches used align with production capacity, manufacturing processes, design flexibility, quality control measures, material knowledge, and integration requirements by considering manufacturing capabilities. This alignment is essential for successful implementation of modular construction and for delivering high-quality modular components or modules.

2.3.5 Construction Process

The construction process can impact a building project's degree of modularity (Olawumi et al., 2022). Modularity may be lower if the construction process is highly fragmented, with many contractors and subcontractors involved. Conversely, if the construction process is highly integrated, with a single contractor or team responsible for all aspects of the project, the degree of modularity may be higher (Zhong et al., 2015). The construction process is a significant determinant of the modularity methodology chosen for a project. The construction process refers to the activities and tasks of constructing a building or structure. Different modularity methodologies can be adopted depending on the specific requirements and characteristics of the construction process (Yin et al., 2019). The construction process can influence the choice of modularity methodology through the on-site versus off-site construction comparison. The construction process can determine whether on-site or off-site construction is more suitable. On-site construction involves traditional methods where most or all construction activities occur directly at the project site (Pan et al., 2018). On the other hand, off-site construction involves manufacturing components or modules in a controlled factory environment and transporting them to the site for assembly. The construction process and project requirements, such as site accessibility, timeline, and site conditions, can influence the decision to choose on-site or off-site construction methods.

Sequence and phasing are other approaches through which the construction process can influence the choice of modularity (Abdelmageed & Zayed, 2020). The sequence and phasing of construction activities can impact the selection of modularity methodology. Modular construction can provide advantages for projects with specific sequencing requirements, such as phased construction or fast-track schedules. Off-site fabrication and assembly of modules allow for parallel construction activities, reducing construction time and improving efficiency (Li et al., 2013). The modularity methodology chosen should align with the desired construction sequence and phasing. The level of integration and coordination required between different trades and construction activities can also influence the choice of modularity methodology. Modular construction involves assembling pre-fabricated components or modules, which requires coordination and integration of various structural, architectural, mechanical, and electrical components. The construction process must consider the extent of integration and coordination required and choose modularity methodology that facilitate seamless assembly and integration of modular elements (Cheung et al., 2002). The construction site's constraints and limitations can be another construction process approach that impacts the choice of modularity methodology. Factors such as site accessibility, space constraints, environmental considerations, or regulatory restrictions can influence the suitability of different modularity methodology (Wang & Hu, 2017). For example, projects with limited site access or challenging conditions may benefit from off-site fabrication and assembly as it minimises disruption and site-related challenges.

Furthermore, the other approaches to the construction process that impact the choice of modularity methodology are construction techniques and practices, as well as skilled labour and workforce considerations. The existing construction techniques and practices employed in the industry can influence the selection of modularity methodology (Han et al., 2012). Some modularity methodologies align more naturally with specific construction techniques and practices, while others may require a shift in the traditional construction approach. To ensure smooth implementation and integration, stakeholders must consider the compatibility between existing construction techniques and the chosen modularity methodology (Taghaddos et al., 2014). The availability and expertise of skilled labour and workforce can impact the choice of modularity methodology. Some modularity methodologies may require specific skills or training for the manufacturing and assembly processes. The construction process should consider the availability of skilled labour and assess whether the required workforce can be readily trained or if additional resources need to be allocated (Arashpour et al., 2015). Stakeholders can select the best modularity methodology by considering the construction process's unique requirements. The selected methods should align with the available labour force, the integration requirements, the site limitations, the construction sequence, and the current construction practices. This alignment ensures a smooth and efficient construction process by optimising the benefits of modularity methodology in the construction industry.

2.3.6 Cost and Time Constraints

Cost and time constraints can influence the degree of modularity in a building project (Azhar et al., 2013a, 2013b). Projects with tight budgets or timelines may require more standardised components to reduce costs and accelerate construction times, which can increase the degree of modularity in the project. Cost and time constraints are essential in selecting the most suitable modularity methodology for a construction project (Abdel-Wahab & Vogl, 2011). These constraints directly impact project budgets, schedules, and overall feasibility. The mechanisms in which cost and time constraints influence the choice of modularity are cost efficiency, time savings, offsite versus onsite work, budget allocation, project scale and complexity, and return on investment (ROI). Cost constraints play a significant role in determining the modularity methodology (Korman & Lu, 2011). Modular construction methods can offer cost advantages over traditional construction approaches. Modular construction frequently saves costs by utilising economies of scale, more straightforward manufacturing procedures, lower labour costs, and less material waste. The extent of cost savings can vary depending on the chosen modularity methodology. For example, off-site modular construction may offer higher cost efficiencies due to optimised manufacturing processes and reduced on-site labour requirements.

Time constraints are another crucial factor in selecting modularity methodology. Modular construction is well-known for its ability to accelerate construction schedules compared to traditional methods (Gibb & Isack, 2003). Modular construction allows for the simultaneous completion of on-site and off-site tasks, such as manufacturing modules off site while site preparation is underway. This can significantly shorten the total construction period. The level of modularity and pre-fabrication in the chosen methodology should align with the project's time constraints to maximise time savings. Cost and time constraints can also influence the balance between off-site and on-site work. Owing to controlled factory environments and optimised manufacturing processes, off-site modular construction typically allows for greater productivity, quality control, and faster production (Pan et al., 2012). However, transportation costs and logistics considerations for delivering modules to the site must be factored in. On-site work, while providing more control over the construction process, may have higher labour costs and longer construction durations. The choice of modularity methodology should strike a balance between off-site and on-site work to optimise cost and time efficiency.

Additionally, the available budget for the project is a critical factor in selecting modularity methodology (Jeng et al., 2011). Depending on the budget constraints, a specific modularity methodology may be more feasible than others. Cost-effective modular construction approaches, such as panelised systems or hybrid methods that combine modular and traditional construction techniques, may be preferred for projects with limited budgets. These methodologies allow for partial modularity while providing cost savings and time benefits. The scale and complexity of the project can also impact the choice of modularity methodology (Diez, 2007). Larger-scale projects often have more opportunities for modularity, as the repetition of components or modules can lead to increased cost savings and efficiency. However, highly customised or complex projects may require a more flexible modularity methodology that can accommodate unique design requirements. The budget and time constraints of the project's scale and complexity should be carefully considered to determine the most suitable modularity approach. Lastly, cost and time constraints should be evaluated in terms of the project's overall return on investment (ROI).

While modularity methodology may offer cost and time benefits, it is essential to assess the long-term value and benefits it brings (Abowitz & Toole, 2010). Factors such as operational efficiencies, maintenance costs, future adaptability, and the project's lifespan should be considered to determine the optimal balance between cost, time, and overall ROI. Considering cost and time constraints, stakeholders can select the most appropriate modularity methodology that aligns with project budgets, schedules, and overall feasibility. Striking the right balance between cost savings, time efficiencies, and project requirements ensures a successful implementation of modular construction while maximising the project's value.

2.4 Limitations of Modular Construction

It is vital to understand that the concept of modular construction, also known as prefabricated or off-site construction, is a building method that involves manufacturing building components off-site and assembling them on the construction site (Adekunle et al., 2024). However, while this method has gained popularity due to its potential benefits, such as increased efficiency and reduced costs, it also has limitations that must be considered. These limitations can affect various aspects of the construction process, including the design, transportation and installation, and the finished product's overall quality and durability (Ribeiro et al., 2022). In this context, it is essential to understand the limitations of modular construction to make informed decisions about its use and ensure that the final product meets the desired standards. Explained below are the identified limitations to the adoption of modular construction.

2.4.1 Perception and Stigma

Technological stigma refers to the negative perception of potential hazards of new technology or technology-based activities such as modular construction (Ehrari et al., 2020). The stigma can result from actual or perceived hazards fuelled by negative imagery from mass media or stakeholders' perceptions. To reduce the risk of innovative technology stigmas, public policy towards technological innovations should promote close interaction with the public. The public can learn about the benefits of adopting new technologies such as modular construction through such interaction. A logical next step is to examine whether a relationship exists between user perceptions of novelty and the challenges associated with evaluating and adopting specific technological innovations (Kubjana et al., 2024). Prior research has outlined some challenges in convincing users to adopt specific technological innovations. For instance, a study by Hossain & Dwivedi (2014) examined consumer perceptions of radio frequency identification (RFID). It concluded that certain risks (e.g., information privacy) must be alleviated before adopting the methodology. Another study by Tambwe et al. (2025) focused on biometrics as a user authentication method. It stressed the need to educate users about the rewards or benefits that would be realised should the methodology be adopted. In both cases, the possibility exists that varying degrees of perceived novelty may leave potential users predisposed to either respective risks or rewards that come with adopting technological innovations. Thus, construction stakeholders should consider whether a potential adopter's perception of novelty is an initial, affective belief about a technological innovation that precedes any cognitive assessment of the risks and rewards of adoption.

2.4.2 Building Codes and Regulations

Building codes ensure acceptable and safe living environments, especially for low-income and unsophisticated apartment tenants (Bello et al., 2023). However, as modular construction redefines what living environments are acceptable and safe, the building codes should adapt to these changes in the construction process. The global modular construction market is estimated to reach over \$53,000 million by 2023. It is projected to register a compound annual growth rate (CAGR) of over 8.5% between 2023 and 2028 (Ismail et al., 2022). Modular buildings will enhance the use of products that promise to provide numerous safety, convenience, and health benefits, such as intelligent sensors. Building codes have two broad purposes: they set minimum requirements for habitability, thus protecting low-income and vulnerable populations from unscrupulous landlords who otherwise would seek to cut costs in construction and maintenance; and they set 'standard practices for the real estate industry, protecting the industry from unfair competition and tenant legal action. Having regulations promoting modular construction is vital, improving living environments by affording adopters' multiple safety and efficiency benefits (Horínkova, 2021). Indeed, at some critical point, the benefits of these modular buildings will likely be significant enough that living environments without the technology will fall below what society establishes as the minimum habitability requirements, reflecting the pervasive spread of this technology. Unfortunately, multifamily apartments, especially those of lower-income residents, may not adopt modular homes as readily as owner-occupied housing because of the economic challenges. Thus, regulation through building codes may be a necessary benefit of modular home technology for low-income multi-family apartment dwellers.

2.4.3 Transportation and Logistics

There are fundamental differences between traditional and modular construction regarding material delivery, coordination, temporary storage and inventory. For example, the modular components are typically large and cumbersome, necessitating extra caution when carrying them across public roads (Azhar et al., 2013a, 2013b). Modular construction methods introduce additional complexity to logistics planning, given the interplay of manufacturing, transportation and assembly processes. Niu et al. (2019) summarised that the logistics service in modular construction features three features: specialisation, high risks and time-driven. Ensuring the just-in-time (JIT) delivery of modules to the site is essential but still challenging to achieve in many cases. Some scholars argued that planning off-site transport of materials requires a systematic approach to accounting for material characteristics and the effects of delays in material delivery on the duration and costs of projects (Ahmadian et al., 2016; Jaśkowski et al., 2018; Yazdani et al., 2021). However, there is a lack of understanding of the planning and control of the transportation phase in modular construction as supply chain management principles seem to be neither widespread nor wholly adopted in construction. Droge et al. (2012) proposed various strategies for ensuring the desired logistics performance regarding module supply. For example, at the early stage of modular building projects, the project team will need a plan for material flow from the factory to the designated site and a mechanism to streamline and manage the process. To optimise the intermodal transportation performance, the coordination among all the involved parties should be improved (Gharehgozli & Zaerpour, 2018). To ensure a smooth installation workflow within the timeframe, a staging area close to the site can help to avoid module delivery delays and to address the limited site storage area for modules. However, preparing such a facility appears to be a challenge to modular projects in various forms.

2.4.4 Limited Design Flexibility

Since building modules are mostly prefabricated in factories, design choices are limited. The modular buildings are designed and fabricated considering safe transportation and assembly (Fard et al., 2015). For example, module size can be limited to the on-road allowable size, i.e., maybe 10 * 70 feet (maximum width * length), and the weight must be within the crane's maximum capacity to lift, i.e., approximately 25 tons. Initially, modular construction may seem straightforward; however, it requires a higher level of expertise to design a modular building and install the modules efficiently (Otasowie et al., 2025). It involves complex planning and expert modular architects, engineers, and contractors to be familiar with the specifications to safely and conveniently fit modules through accurate interconnections. During the designing stage, buyers and owners have to decide on the final selection of finishes and appliances to prefabricate modules. Generally, once a modular project is ready to proceed beyond schematic design, the modular manufacturer should be brought on board to provide consultation and, potentially, design-assist services (Ferdous et al., 2019). The earlier this happens, the better, especially for project teams new to modular construction. Having the modular manufacturer involved early allows the project team to benefit from the manufacturer's expertise and knowledge of available products and techniques. In most cases, because structural design is one of the most critical elements of modular construction, the modular manufacturer will take on the structural engineering package or, at a minimum, provide structural design-assist services (Fifield et al., 2018). As modular manufacturers are generally more experienced with the complexities of modular construction, they can better check that related efficiencies are gained and associated risks are addressed during the design phase. Architects should understand that this involves a fundamental shift in traditional design responsibilities.

2.5 Summary

From the literature reviewed, the lesson from the topics covered highlights a comprehensive understanding of modularity in construction methodology. The introduction establishes the concept, purpose, and relevance of modular approaches in modern construction, setting the stage for deeper exploration. The forms of modularity construction methodology demonstrate the different ways modular systems can be applied, ranging from volumetric modules to panelized systems, each with distinct advantages and suitability for specific project needs. The requirements section underscores the technical, logistical, and regulatory prerequisites necessary for successful implementation, including precision in design, manufacturing standards, and efficient site coordination. Lastly, the limitations section identifies the challenges, such as high initial costs, transportation difficulties, and industry resistance, which must be addressed to maximise adoption. Overall, the lesson emphasises that while modular construction offers efficiency, quality, and sustainability benefits, its success depends on meeting essential requirements and effectively mitigating inherent limitations.

References

Abdelmageed, S., & Zayed, T. (2020). A study of literature in modular integrated construction—Critical review and future directions. *Journal of Cleaner Production, 277*, 124044. https://doi.org/10.1016/j.jclepro.2020.124044

Abdel-Wahab, M., & Vogl, B. (2011). Trends of productivity growth in the construction industry across Europe, US and Japan. *Construction Management and Economics, 29*(6), 635–644.

Abowitz, D. A., & Toole, T. M. (2010). Mixed method research: Fundamental issues of design, validity, and reliability in construction research. *Journal of Construction Engineering and Management, 136*(1), 108–116. https://doi.org/10.1061/(asce)co.1943-7862.0000026

Adekunle, P., Aigbavboa, C., Otasowie, K., & Akinradewo, O. (2024). Matching-up modularity methodology application within the built environment: A bibliometric review. *Journal of Asian Architecture and Building Engineering*, 1–16. https://doi.org/10.1080/13467581.2024.2368653

Ahmadian, F. A., Akbarnezhad, A., Rashidi, T. H., & Waller, S. T. (2016). Accounting for transport times in planning off-site shipment of construction materials. *Journal of Construction Engineering and Management, 142*(1), 04015050.

Alvanchi, A., Azimi, R., Lee, S., AbouRizk, S. M., & Zubick, P. (2012). Off-site construction planning using discrete event simulation. *Journal of Architectural Engineering, 18*(2), 114–122. https://doi.org/10.1061/(asce)ae.1943-5568.0000055

Antonini, E., Longo, D., & Gianfrate, V. (2014). Towards nZEB: Modular pre-assembled steel systems for residential buildings. *WIT Transactions on the Built Environment, 142*, 349–360. https://doi.org/10.2495/ARC140301

Arashpour, M., Wakefield, R., Blismas, N., & Maqsood, T. (2015). Autonomous production tracking for augmenting output in off-site construction. *Automation in Construction, 53*, 13–21. https://doi.org/10.1016/j.autcon.2015.03.013

Ayinla, K. O., Cheung, F., & Tawil, A. R. (2020). Demystifying the concept of off-site manufacturing method: Towards a robust definition and classification system. *Construction Innovation, 20*(2), 223–246. https://doi.org/10.1108/CI-07-2019-0064

Azhar, S., Lukkad, M. Y., & Ahmad, I. (2013a). An investigation of critical factors and constraints for selecting modular construction over conventional stick-built technique. *International Journal of Construction Education and Research, 9*(3), 203–225. https://doi.org/10.1080/15578771.2012.723115

Azhar, S., Lukkad, M. Y., & Ahmad, I. (2013b). An investigation of critical factors and constraints for selecting modular construction over conventional stick-built technique. *International Journal of Construction Education and Research, 9*(3), 203–225. https://doi.org/10.1080/15578771.2012.723115

Baldassano, S. N., & Bassett, D. S. (2016). Topological distortion and reorganized modular structure of gut microbial co-occurrence networks in inflammatory bowel disease. *Scientific Reports, 6*, 1–14. https://doi.org/10.1038/srep26087

Baldea, M., Edgar, T. F., Stanley, B. L., & Kiss, A. A. (2017). Modular manufacturing processes: Status, challenges, and opportunities. *AIChE Journal, 63*(10), 4262–4272. https://doi.org/10.1002/aic.15872

Bastian, Z., Schnieders, J., Conner, W., Kaufmann, B., Lepp, L., Norwood, Z., Simmonds, A., & Theoboldt, I. (2022). Retrofit with passive house components. In *Energy Efficiency*, Vol. 15, Issue 1. Netherlands. https://doi.org/10.1007/s12053-021-10008-7

Begic, H., & Krstic, H. (2022). Possibilities of using prefabricated modular panels for building NZEB buildings in earthquake-affected areas in Croatia—Case study. *International Journal of Sustainable Construction Engineering and Technology, 13*(1), 137–148. https://publisher.uthm.edu.my/ojs/index.php/IJSCET/article/view/10321

Bello, A. O., Eje, D. O., Idris, A., Semiu, M. A., & Khan, A. A. (2023). Drivers for the implementation of modular construction systems in the AEC industry of developing countries. *Journal of Engineering, Design and Technology, 2018*. https://doi.org/10.1108/JEDT-11-2022-0571

Boafo, F. E., Kim, J. H., & Kim, J. T. (2016). Performance of modular prefabricated architecture: Case study-based review and future pathways. *Sustainability (Switzerland), 8*(6), 1–16. https://doi.org/10.3390/su8060558

Boyd, N., Khalfan, M. M. A., & Maqsood, T. (2013). Off-site construction of apartment buildings. *Journal of Architectural Engineering, 19*(1), 51–57. https://doi.org/10.1061/(asce)ae.1943-5568.0000091

Browning, T. R., & Yassine, A. A. (2010). Resource-constrained multi-project scheduling: Priority rule performance revisited. *International Journal of Production Economics, 126*(2), 212–228. https://doi.org/10.1016/j.ijpe.2010.03.009

Chen, Z., Liu, J., Yu, Y., Zhou, C., & Yan, R. (2017). Experimental study of an innovative modular steel building connection. *Journal of Constructional Steel Research, 139*, 69–82. https://doi.org/10.1016/j.jcsr.2017.09.008

Cheung, S. O., Tong, T. K. L., & Tam, C. M. (2002). Site pre-cast yard layout arrangement through genetic algorithms. *Automation in Construction, 11*, 35–46. https://doi.org/10.1016/S0926-5805(01)00044-9

D'Agostino, D., Tzeiranaki, S. T., Zangheri, P., & Bertoldi, P. (2021). Assessing nearly zero energy buildings (NZEBs) development in Europe. *Energy Strategy Reviews, 36*, Article 100680. https://doi.org/10.1016/j.esr.2021.100680

Deng, E., Zong, L., Ding, Y., Zhang, Z., Zhang, J., & Shi, F. (2020). Since January 2020 Elsevier has created a COVID-19 resource centre with free information in English and Mandarin on the novel coronavirus COVID-19. The COVID-19 resource centre is hosted on Elsevier Connect, the company's public news and information.

Diez, R. (2007). AUTMOD3: The integration of design and planning tools for automatic modular construction. *International Journal of Advanced Robotic Systems, 4*(4), 457–468.

Droge, C., Vickery, S. K., & Jacobs, M. A. (2012). Does supply chain integration mediate the relationships between product/process strategy and service performance? An empirical study. *International Journal of Production Economics, 137*(2), 250–262. https://doi.org/10.1016/j.ijpe.2012.02.005

Economidou, M., Todeschi, V., Bertoldi, P., D'Agostino, D., Zangheri, P., & Castellazzi, L. (2020). Review of 50 years of EU energy efficiency policies for buildings. *Energy and Buildings, 225*, Article 110322. https://doi.org/10.1016/j.enbuild.2020.110322

Ehrari, H., Ulrich, F., & Andersen, H. B. (2020). Concerns and trade-offs in information technology acceptance: The balance between the requirement for privacy and the desire for safety. *Communications of the Association for Information Systems, 47*(1), 227–247. https://doi.org/10.17705/1cais.04711

Fard, M. M., Terouhid, S. A., Kibert, C. J., & Hakim, H. (2015). Safety concerns related to modular/prefabricated building construction. *International Journal of Injury Control and Safety Promotion, 24*(1), 10–23. https://doi.org/10.1080/17457300.2015.1047865

Ferdous, W., Bai, Y., Ngo, T. D., Manalo, A., & Mendis, P. (2019). New advancements, chal-lenges and opportunities of multi-storey modular buildings—a state-of-the-art review. *Engineering Structures, 183*, 883–893. https://doi.org/10.1016/j.engstruct.2019.01.061

Fifield, L., Lomas, K., Giridharan, R., & Allinson, D. (2018). Hospital wards and modular construction: Summertime overheating and energy efficiency. *Building and Environment, 141*, 28–44. https://doi.org/10.1016/j.buildenv.2018.05.041

Gbadamosi, A. Q., Mahamadu, A. M., Oyedele, L. O., Akinade, O. O., Manu, P., Mahdjoubi, L., & Aigbavboa, C. (2019). Offsite construction: Developing a BIM-Based optimizer for assembly. *Journal of Cleaner Production, 215*, 1180–1190. https://doi.org/10.1016/j.jclepro.2019.01.113

Gharehgozli, A., & Zaerpour, N. (2018). Stacking outbound barge containers in an automated deep-sea terminal. *European Journal of Operational Research, 267*(3), 977–995. https://doi.org/10.1016/j.ejor.2017.12.040

Gibb, A. G. F., & Isack, F. (2003). Re-engineering through pre-assembly: Client expectations and drivers. *Building Research and Information, 31*(2), 146–160.

Ginigaddara, B., Perera, S., Feng, Y., & Rahnamayiezekavat, P. (2019). Typologies of offsite construction. In *World Construction Symposium* (pp. 567–577). https://doi.org/10.31705/WCS.2019.56

Ginigaddara, B., Perera, S., Feng, Y., & Rahnamayiezekavat, P. (2022). Development of an offsite construction typology: A delphi study. *Buildings, 12*(1), 20. https://doi.org/10.3390/buildings12010020

Hammad, A. W., Akbarnezhad, A., Wu, P., Wang, X., & Haddad, A. (2019). Building information modelling-based framework to contrast conventional and modular construction methods through selected sustainability factors. *Journal of Cleaner Production, 228*, 1264–1281. https://doi.org/10.1016/j.jclepro.2019.04.150

Han, S. H., Al-Hussein, M., Al-Jibouri, S., & Yu, H. (2012). Automated post-simulation visualization of modular building production assembly line. *Automation in Construction, 21*, 229–236. https://doi.org/10.1016/j.autcon.2011.06.007

Hong, J., Shen, G. Q., Li, Z., Zhang, B., & Zhang, W. (2018). Barriers to promoting prefabricated construction in China: A cost–benefit analysis. *Journal of Cleaner Production, 172*, 649–660. https://doi.org/10.1016/j.jclepro.2017.10.171

Horínkova, D. (2021). Advantages and disadvantages of modular construction, including environmental impacts. *IOP Conference Series: Materials Science and Engineering, 1203*(3), 32002.

Hossain, M. A., & Dwivedi, Y. K. (2014). What improves citizens' privacy perceptions toward RFID technology? A cross-country investigation using mixed method approach. *International Journal of Information Management, 34*(6), 711–719. https://doi.org/10.1016/j.ijinfomgt.2014.07.002

Hsu, P. Y., Angeloudis, P., & Aurisicchio, M. (2018). Optimal logistics planning for modular construction using two-stage stochastic programming. *Automation in Construction, 94*, 47–61. https://doi.org/10.1016/j.autcon.2018.05.029

Hyun, H., Kim, H., Lee, H. S., Park, M., & Lee, J. (2020). Integrated design process for modular construction projects to reduce rework. *Sustainability (Switzerland), 12*(2), 1–19. https://doi.org/10.3390/su12020530

Innella, F., Arashpour, M., & Bai, Y. (2019). Lean methodologies and techniques for modular construction: Chronological and critical review. *Journal of Construction Engineering and Management, 145*(12). https://doi.org/10.1061/(asce)co.1943-7862.0001712

Isaac, S., Bock, T., & Stoliar, Y. (2016). A methodology for the optimal modularisation of building design. *Automation in Construction, 65*, 116–124. https://doi.org/10.1016/j.autcon.2015.12.017

Ismail, S., Hon, C. K. H., Crowther, P., Skitmore, M., & Lamari, F. (2022). The drivers and challenges of adopting the Malaysia industrialised building system for sustainable infrastructure development. *Construction Innovation.* https://doi.org/10.1108/CI-05-2021-0088

Jaśkowski, P., Sobotka, A., & Czarnigowska, A. (2018). Decision model for planning material supply channels in construction. *Automation in Construction, 90*, 235–242. https://doi.org/10.1016/j.autcon.2018.02.026

Jeng, B., DiGiovanni, D., & Wan, A. (2011). High-performance modular building: Inspiration from the past, technology from the present, design for the future. In *Proceedings of the 2011 Architectural Engineering National Conference*, Oakland, CA, 2011.

Kamali, M., & Hewage, K. (2017). Development of performance criteria for sustainability evaluation of modular versus conventional construction methods. *Journal of Cleaner Production, 142*, 3592–3606. https://doi.org/10.1016/j.jclepro.2016.10.108

Kim, J.-Y., & Lee, J. K. (2014). Abasic study on the application of modular construction—focused on the analysis of case study. *Journal of the Korean Housing Association, 25*(4), 39–46. https://doi.org/10.6107/JKHA

Korman, T. M., & Lu, N. (2011). Innovation and improvements of mechanical, electrical, and plumbing systems for modular construction using building information modelling. In *Proceedings of the 2011 Architectural Engineering National Conference*, Oakland, CA, 2011.

Kubjana, L., Adekunle, P., Aigbavboa, C., & Otasowie, K. (2024). Digitisation in the South African construction industry: Recurring challenges. In: K. Arai (Ed.), *Proceedings of the Future Technologies Conference (FTC) 2024*, Volume 4. FTC 2024. Lecture Notes in Networks and Systems, Vol. 1157. Cham. https://doi.org/10.1007/978-3-031-73128-0_35

Lawson, R. M., & Ogden, R. G. (2008). "Hybrid" light steel panel and modular systems. *Thin-Walled Structures, 46*(7–9), 720–730. https://doi.org/10.1016/j.tws.2008.01.042

Lee, J., & Hyun, H. (2019). Multiple modular building construction project scheduling using genetic algorithms. *Journal of Construction Engineering and Management, 145*(1), 1–11. https://doi.org/10.1061/(asce)co.1943-7862.0001585

Lehmann, S. (2012). Sustainable construction for urban infill development using engineered massive wood panel systems. In *Sustainability*, Vol. 4, Issue 10. https://doi.org/10.3390/su4102707

Li, H. X., Al-Hussein, M., Lei, Z., & Ajweh, Z. (2013). Risk identification and assessment of modular construction utilizing fuzzy analytic hierarchy process (AHP) and simulation. *Canadian Journal of Civil Engineering, 40*(12), 1184–1195. https://doi.org/10.1139/cjce-2013-0013

Liu, S., & Qian, S. (2019). Evaluation of social life-cycle performance of buildings: Theoretical framework and impact assessment approach. *Journal of Cleaner Production, 213*, 792–807. https://doi.org/10.1016/j.jclepro.2018.12.200

Mao, C., Shen, Q., Shen, L., & Tang, L. (2013). Comparative study of greenhouse gas emissions between off-site prefabrication and conventional construction methods: Two case studies of residential projects. *Energy and Buildings, 66*, 165–176. https://doi.org/10.1016/j.enbuild.2013.07.033

Moghadam, M., Al-Hussein, M., Al-Jibouri, S., & Telyas, A. (2012). Post simulation visualisation model for effective scheduling of modular building construction. *Canadian Journal of Civil Engineering, 39*(9), 1053–1061. https://doi.org/10.1139/l2012-077

Niu, S., Yang, Y., & Pan, W. (2019). Logistics planning and visualization of modular integrated construction projects based on BIM-GIS integration and vehicle routing algorithm. In *Modular and Offsite Construction (MOC) Summit Proceedings* (pp. 579–586). https://doi.org/10.29173/mocs141

Ogunbiyi, O., Oladapo, A., & Goulding, J. (2014). An empirical study of the impact of lean construction techniques on sustainable construction in the UK. *Construction Innovation, 14*(1), 88–107. https://doi.org/10.1108/CI-08-2012-0045

Olawumi, T. O., Chan, D. W. M., Ojo, S., & Yam, M. C. H. (2022). Automating the modular construction process: A review of digital technologies and future directions with blockchain technology. *Journal of Building Engineering, 46*, 103720. https://doi.org/10.1016/j.jobe.2021.103720

Otasowie, K., Aigbavboa, C., Adekunle, P., Ikuabe, M. (2024b). Robotics adoption in the construction sector: A bibliometric analysis of current research trends and future directions. In: K. Arai (Ed.), *Proceedings of the Future Technologies Conference (FTC) 2024, Volume 4*. FTC 2024. Lecture Notes in Networks and Systems, Vol. 1157. Cham. https://doi.org/10.1007/978-3-031-73128-0_41

Otasowie, K., Aigbavboa, C., Oke, A., Adekunle, P., & Zanele, B. (2023). Exploring the drivers of blockchain technology adoption in construction project logistics: A case of South Africa. In: K. Arai (Ed.), *Proceedings of the Future Technologies Conference (FTC) 2023, Volume 3*. FTC 2023. Lecture Notes in Networks and Systems, Vol. 815. Cham. https://doi.org/10.1007/978-3-031-47457-6_22

Otasowie, K., Aigbavboa, C., Oke, A., Ikuabe, M., & Adekunle, P. (2025). Streamlining construction project logistics in developing nations: Making a case for blockchain technology adoption. In: A. GhaffarianHoseini, A. Ghaffarianhoseini, F. Rahimian, & M. Babu Purushothaman (Eds.), *Proceedings of the International Conference on Smart and Sustainable Built Environment* (SASBE 2024). SASBE 2024. Lecture Notes in Civil Engineering, Vol. 591. Singapore. https://doi.org/10.1007/978-981-96-4051-5_134

Otasowie, K., Aigbavboa, C., Oke, A., Adekunle, P., & Liphadzi, M. (2024a). Constraints to clients' acceptance of recycled construction materials in developing nations: Professionals' perspective. *Engineering Proceedings, 76*(1), 60. https://doi.org/10.3390/engproc2024076060

Pan W., Yang, Y., Yang, L. (2018). High-rise modular building: Ten-year journey and future development. In Y. Lee, R. Harris, C. Wang, C. Harper, & C. Berryman (Eds.), *Construction Research Congress 2018* (pp. 523–532). (ASCE) American Society of Civil Engineers. https://doi.org/10.1061/9780784481301.052

Pan, W., Dainty, A. R. J., & Gibb, A. G. F. (2012). Establishing and weighting decision criteria for building system selection in housing construction. *Journal of Construction Engineering and Management, ASCE*. https://doi.org/10.1061/(ASCE)CO.1943-7862.0000543

Patil, G., Choudhury, D., & Mondal, A. (2021). Three-dimensional soil–foundation–superstructure interaction analysis of nuclear building supported by combined piled-raft system. *International Journal of Geomechanics, 21*(4), 1–16. https://doi.org/10.1061/(asce)gm.1943-5622.0001956

Qi, B., Razkenari, M., Costin, A., Kibert, C., & Fu, M. (2021). A systematic review of emerging technologies in industrialized construction. *Journal of Building Engineering, 39*, Article 102265. https://doi.org/10.1016/j.jobe.2021.102265

Quale, J., Eckelman, M. J., Williams, K. W., Sloditskie, G., & Zimmerman, J. B. (2012). Construction matters: Comparing environmental impacts of building modular and conventional homes in the United States. *Journal of Industrial Ecology, 16*(2), 243–253. https://doi.org/10.1111/j.1530-9290.2011.00424.x

Ribeiro, A. M., Arantes, A., & Cruz, C. O. (2022). Barriers to the adoption of modular construction in Portugal: An interpretive structural modeling approach. *Buildings, 12*(10), 1509. https://doi.org/10.3390/buildings12101509

Risholt, B., Time, B., & Hestnes, A. G. (2013). Sustainability assessment of nearly zero energy renovation of dwellings based on energy, economy and home quality indicators. *Energy and Buildings, 60*, 217–224. https://doi.org/10.1016/j.enbuild.2012.12.017

Sanches, R., Mercan, O., & Roberts, B. (2018). Experimental investigations of vertical post-tensioned connection for modular steel structures. *Engineering Structures, 175*, 776–789. https://doi.org/10.1016/j.engstruct.2018.08.049

Shahpari, M., Saradj, F. M., Pishvaee, M. S., & Piri, S. (2020). Assessing the productivity of prefabricated and in-situ construction systems using hybrid multi-criteria decision making method. *Journal of Building Engineering, 27*, 100979. https://doi.org/10.1016/j.jobe.2019.100979

Sharafi, P., Rashidi, M., Samali, B., Ronagh, H., & Mortazavi, M. (2018). Identification of factors and decision analysis of the level of modularization in building construction. *Journal of Architectural Engineering, 24*(2), 1–12. https://doi.org/10.1061/(asce)ae.1943-5568.0000313

Shin, J., & Choi, B. (2022). Design and implementation of quality information management system for modular construction factory. *Buildings, 12*(5). https://doi.org/10.3390/buildings12050654

Taghaddos, H., Hermann, U., AbouRizk, S., & Mohamed, Y. (2014). Simulation-based multigent approach for scheduling modular construction. *Journal of Computing in Civil Engineering, 28*(2), 263–274. https://doi.org/10.1061/(asce)cp.1943-5487.0000262

Tambwe, O. T., Aigbavboa, C. O., Akinradewo, O. I., & Adekunle, P. A. (2025). Measures to address cyber-attacks in construction project data management processes: A cybersecurity perspective. *IET Information Security, 2025*(1). https://doi.org/10.1049/ise2/7398742

Tasan, S. O., & Gen, M. (2008). A priority-based genetic algorithm approach for solving multiple alternative project scheduling problems with resource constraints and variable activity times. In *Conference Proceedings—IEEE International Conference on Systems, Man and Cybernetics* (pp. 2537–2542). https://doi.org/10.1109/ICSMC.2008.4811677

Tavares, V., Lacerda, N., & Freire, F. (2019). Embodied energy and greenhouse gas emissions analysis of a prefabricated modular house: The "Moby" case study. *Journal of Cleaner Production, 212*, 1044–1053. https://doi.org/10.1016/j.jclepro.2018.12.028

Thai, H., Ngo, T., & Uy, B. (2020). A review on modular construction for high-rise buildings. *28*, 1265–1290.

Turner, C., Oyekan, J., & Stergioulas, L. K. (2021). Distributed manufacturing: A new digital framework for sustainable modular construction. *Sustainability (Switzerland), 13*(3), 1–16. https://doi.org/10.3390/su13031515

Tzourmakliotou, D. (2021). Modular disruption in construction industry—the environmental benefits. *Journal of Civil Engineering and Architecture, 15*(6), 318–329. https://doi.org/10.17265/1934-7359/2021.06.004

Wang, Z., & Hu, H. (2017). Improved precast production-scheduling model considering the whole supply chain. *Journal of Computing in Civil Engineering, 31*. https://doi.org/10.1061/(ASCE)CP.1943-5487.0000667

Williamson, M., Ganah, A., & John, G. A. (2019). Barriers to adopting modern methods of construction in the UK. *Journal of Construction Engineering, 2*(1), 30–39.

Wuni, I. Y., & Shen, G. Q. (2020). Barriers to the adoption of modular integrated construction: Systematic review and meta-analysis, integrated conceptual framework, and strategies. *Journal of Cleaner Production, 249*, Article 119347. https://doi.org/10.1016/j.jclepro.2019.119347

Xu, Z., Zayed, T., & Niu, Y. (2020). Comparative analysis of modular construction practices in mainland China, Hong Kong and Singapore. *Journal of Cleaner Production, 245*, Article 118861. https://doi.org/10.1016/j.jclepro.2019.118861

Yakimchuk, T., Linhares, P., & Hermo, V. (2023). Evaluation of a modular construction system in accordance with the Passivhaus standard for components. *Journal of Building Engineering, 76*, Article 107234. https://doi.org/10.1016/j.jobe.2023.107234

Yazdani, M., Gonzalez, E. D. R. S., & Chatterjee, P. (2021). A multi-criteria decision-making framework for agriculture supply chain risk management under a circular economy context. *Management Decision, 59*(8), 1801–1826. https://doi.org/10.1108/MD-10-2018-1088

Yin, X., Liu, H., Chen, Y., & Al-Hussein, M. (2019). Building information modelling for off-site construction: Review and future directions. *Automation in Construction, 101*, 72–91. https://doi.org/10.1016/j.autcon.2019.01.010

Zhang, Y., Wang, J., Hu, F., & Wang, Y. (2017). Comparison of evaluation standards for green building in China, Britain, United States. *Renewable and Sustainable Energy Reviews, 68*, 262–271. https://doi.org/10.1016/j.rser.2016.09.139

Zhong, R. Y., Peng, Y., Fang, J., Xu, G., Xue, F., Zou, W., & Huang, G. Q. (2015). Towards Physical Internet-enabled prefabricated housing construction in Hong Kong. *IFACPapersOnline, 48*, 1079–1086. https://doi.org/10.1016/j.ifacol.2015.06.227

Chapter 3
Theoretical Perspectives on Modularity Methodology

Abstract This chapter explores the theoretical perspectives that underpin modularity methodology in construction, providing a foundation for understanding its conceptual and practical applications. It begins by defining modularity methodology as a systematic approach to designing, manufacturing, and assembling components in a way that enhances efficiency, flexibility, and quality. The chapter highlights the relevance of theoretical perspectives in shaping the adoption and adaptation of modular construction in diverse project contexts. Various approaches to modularity methodology are examined, reflecting how different schools of thought address design principles, process integration, and supply chain coordination. These approaches reveal the multifaceted nature of modularity, where concepts from manufacturing, lean construction, and systems thinking converge to create streamlined and adaptable processes. Furthermore, the discussion includes established modularity methodology models, each providing structured frameworks for planning, execution, and evaluation. These models guide practitioners in aligning design decisions, production strategies, and assembly techniques to achieve desired project outcomes. The chapter highlights how modularity methodology can address important construction challenges like time constraints, quality assurance, cost management, and sustainability goals by integrating these theoretical viewpoints. It also emphasises how crucial it is to modify these frameworks to fit project settings, legal requirements, and stakeholder requirements. In summary, the chapter offers modularity methodology as a flexible, theory-based approach that blends creativity and pragmatism, rather than just as a technical procedure. This theoretical foundation lays the groundwork for future discussions on its useful applications, advantages, and drawbacks in the construction sector and serves as the foundation for well-informed decision-making.

3.1 Introduction

The term 'modularity' is widely used in studies of technological and organisational systems (Campagnolo & Camuffo, 2010). Product systems are deemed 'modular', for example, when they can be decomposed into several components that may be

P. Adekunle et al., *Modularity Methodology*,
https://doi.org/10.1007/978-3-032-12918-5_3

mixed and matched in various configurations. Components can connect, interact, or exchange resources (such as energy or data) in some way by adhering to a standardised interface (Adekunle et al., 2023a, 2023b). Unlike a tightly integrated product, whereby each component is designed to work specifically (and often exclusively) with other components in a tightly coupled system, modular products are systems of loosely coupled components.

Organisational systems become increasingly modular when substituting loosely coupled forms for tightly integrated, hierarchical structures (O'Connor et al., 2015). For instance, when the firm utilises contract manufacturing rather than in-house manufacturing, it uses an organisational component more independently than building such capabilities in-house: the firm can switch between contract manufacturers that perform different functions, and the contract manufacturer can similarly work for different firms. As firms in a given industry begin to substitute loose coupling with organisational components outside of firm boundaries for activities that were once conducted in-house, the entire production system (which may encompass many firms) becomes increasingly modular (Akhavian & Behzadan, 2014). The firms themselves become more specialised components. Using loosely coupled structures enables firms to achieve greater flexibility in both scope and scale. This aligns with modularity in production processes, which relates to how technological artefacts are produced. This consists of the artefact's entire value chain, from the design of the artefact to the manufacturing and distribution stages. Production emphasises the separation of concerns and the creation of discrete and reusable components (Otasowie et al., 2024). Therefore, the firm can easily switch between different providers of these activities (e.g., between different contract manufacturers or alliance partners) compared to building the capabilities for all activities in-house, thus responding to different market needs more quickly. However, these flexibility gains come with a price. Therefore, the organisation must assess each form's achievable flexibility gains and any accompanying performance loss.

3.2 Approaches and Perspectives on Modularity Methodology

According to Acebes et al. (2015), modularisation within firms leads to the disaggregation of the traditional form of hierarchical governance. The firm is decomposed into relatively small autonomous organisational units (modules) to reduce complexity. Modularisation leads to a structure where the modules integrate strongly interdependent tasks while the interdependencies between the modules are weak. In this connection, the dissemination of modular organisational forms has been facilitated by the widespread efforts of most large firms to re-engineer, refocus and restructure (Haas & Fagerlund, 2002). These efforts usually involve a strong process orientation: the complete service-provision process of the business is split up into partial processes, which can then be handled autonomously by cross-functional teams within organisational units (modules). The coordination of the modules is often carried out by using internal market mechanisms, in particular by the implementation of profit centres. Modularisation enables a more flexible and quicker reaction to changing general or market conditions (Kalantarnia et al., 2010).

Building on the above principles, many alternative forms of modularisation of organisations (for-profit or non-profit) are possible. However, it is crucial to note that modularisation is not an independent and self-contained organisational concept but consists of several essential ideas integral to other organisational concepts (Meel & Seider, 2006). As discussed so far, it is evident that the study of modularity in all spheres of life is to build a system in modules. Hence, this shows the relevance of this study in determining factors that influence modularity methodology in the construction industry. This ensures that modular construction is adopted to enhance construction processes and attain project completion within the shortest possible time.

Additionally, modularity has been widely discussed and debated in various fields, including biology, engineering, computer science, psychology, and philosophy. Modularity is dividing a system or object into distinct, interdependent parts or modules that can function independently or in coordination (Choi et al., 2016). In recent years, there has been a growing interest in exploring the nature of modularity, its implications, and the different theories that attempt to explain it. In this research, the discussions delve into the theories on modularity, examining how these theories explain the formation, structure, and evolution of modular systems. There is a discussion of the concept of modularity and its significance in the construction domain. The different theoretical perspectives on modularity, including structuralist, functionalist, and computational theories, are also reviewed for analysis of the strengths and weaknesses of each theory and their contribution to the understanding of modularity. Furthermore, modularity is manifested in different systems, such as biological organisms, technological devices, and cognitive processes (Naqvi et al., 2014). The advantages and disadvantages of modularity in construction affect system performance, adaptability, and robustness. The current study conforms to the later conceptualisation as the various methodologies of modularity are moderated by different variables which influence the adoption of modular construction.

Notably, the definition of modularity is based on the concept of conditional independence, described in previous sections, and the analytical recognition of patterns of conditional independence among sets of traits (Pasquire et al., 2004). The definition applies to both continuous and discrete traits of modularity. A module is a maximal subset of traits for which all pairs of traits within the subset are mutually informative and conditional on all other traits under consideration. The term 'informative' is used in the sense of information theory. According to Ellerman (2018), classical information theory is concerned with measuring the information in one random variable in relation to the value of another. Information adequately related to the prediction problem of 'average predictability'. The insistence on conditional informativeness results from the observation that, for most datasets, there are generally high correlations among traits. Much of this correlation may be due to joint associations with other variables, either observed or unobserved. By focusing on conditional association, the definition emphasises interactions that provide the most significant amount of information about patterns of variability while ignoring those interactions that contribute little or nothing to further knowledge about the systems under study (Nadim & Goulding, 2011). As mentioned earlier, the definition of modularity has a precise interpretation in the context of an independence graph based on covariances or correlations among the traits.

Over the years, several authors have used some form of comparison to model modularity. Some theoretical approaches that have been advanced, among others, include the lean production theory, the theory of interdependence and modularity, the queuing theory, the control theory, the social cognitive theory, and the technology-acceptance theory, among others.

3.2.1 Lean Production Theory (2019)

Lean is a process improvement methodology that delivers better, faster, cheaper products and services. Kholopane and Vanadayar (2014) defined lean as a way to specify value, line up value-creating actions in the best sequence, conduct those activities without interruption whenever someone requests them, and perform them more and more effectively. In short, lean thinking is lean because it provides a way to do more and more with less human effort, less human equipment, less time, and less space while coming closer and closer to providing clients with exactly what they want. Lean principles can be widely applied to improve the productivity and efficiency of modular construction operations, while simulation augments lean theory by allowing its benefits and issues to be analysed quantitatively before actual implementation (Blismas & Wakefield, 2009). Lean production is a philosophy that incorporates certain ideologies and principles into the design, completion, and delivery of a modular construction project. This theory aims to shape how a modular construction project is handled through systemised thought and action, reducing effort and products, maximising efficiency and profit, and maximising the output and overall value of the project. Lean construction adopts the principles and practices of lean production and applies them to modular building processes (Pan & Goodier, 2012). It uses operational research and practical development when designing the project, focusing on continuous improvement throughout construction. That includes everything from environmental design to maintenance and salvaging. Lean recommendations include implementing total quality management (TQM) to reduce the presence of defects, an Internet-based system to optimise deliveries, minimise cycle times, cross-training to reduce idle time of resources and using construction robotics to replace the need for manual labour when carrying out modular construction (Eastman & Sacks, 2008). Lean construction and simulation have been used to improve construction processes such as housebuilding, bridge deck construction and block and bricklaying operations. Lean construction aims to improve modular construction operations by minimising waste and maximising value. Applying lean construction principles to modular construction can increase efficiency and promote adoption. Simulation augments lean construction by allowing its benefits and issues to be evaluated and understood quantitatively before implementation, thereby aiding decision-making capabilities from the managerial perspective (Khalfan & Maqsood, 2014). Lean production, which has its roots in the manufacturing industry, applies to construction due to similar goals. Lean construction continuously seeks opportunities to reduce waste and maximise value in modular construction projects. However, Saurin et al. (2013) argued that although lean theory is a powerful method for process improvement, it is deterministic and therefore unable to address variability in a system or analyse the performance of a future system.

3.2.2 Theory of Interdependence and Modularity (2019)

The theory of interdependence and modularity (also known as the modularity theory) explains how different parts of a product's architecture relate to one another and consequently affect metrics of production and adoption (Kubota et al., 2017). The study explained that a product is modular when there are no unpredictable elements in the design of its parts. Modularity standardises how components fit together physically, mechanically, and chemically (Gentile, 2013). The parts fit and work together in well-understood, crisply codified ways. A product is interdependent when one part is made and delivered depending on how others are made and delivered. Interdependency between parts requires the same organisation to develop both components if it hopes to develop either. Examining these kinds of real-world accomplishments reveals that modularity has multiple facets. These dimensions together achieve an instantiation of a modular entity within an architectural construct composed of discrete functional building blocks at different levels of abstraction, being brought together in an ordered connectivity (Oyama et al., 2010).

To construct a mathematical representation of modularity, the analysis starts with an analytic assignment of functional attributes to the observed modular parameters of diversity, modular replication, and interface fit. Modular diversity accounts for the different types of functional modules brought forward for entity agglomeration. Modular replication accounts for the replication of the types of functional modules brought forward for entity agglomeration. In formulating and testing an equation to describe modularity, it became apparent that certain aspects of earlier papers had initiated methods to describe modular construction. Mikkola (2007) developed a composite measure of modularity, taking a network approach to describe a modular system mathematically. It should be recognised that many modular constructs provide insight into the natural processes involved in modularity. When viewed as modular entities, chemistry, theoretical physics, and particle physics reflect the similar modular composition and combinatorial structure discussed above. It is this underlying commonality that suggests the existence of a modular propensity in nature (Molavi & Barral, 2016). In addition, other fractal investigations display high levels of modularity. This evidence suggests a general mathematical representation of a modular entity's existence.

3.2.3 *Queuing Theory (2013)*

Queuing theory is a branch of mathematics that studies how lines form, how they function, and why they malfunction (Lakshmi et al., 2013). Queuing theory examines every component of waiting in line, including the arrival process, service process, number of servers, number of system places, and the number of customers, which might be people, data packets, cars, or anything else. Real-life applications of queuing theory cover various businesses, including modular construction. Its findings may be used to provide faster customer service, increase traffic flow, improve order shipments from a warehouse, or design data networks and call centres. As a branch of operations research, queuing theory can help inform construction business decisions on how to build more efficient and cost-effective workflow systems (Mehandiratta, 2011). Queues can occur whenever resources are restricted. Some queuing is tolerable in any business since a total absence of a queue would suggest a costly overcapacity. Queuing theory aims to design balanced systems that aid modular construction sequences quickly and efficiently but do not cost too much to be sustainable. At its most basic level, queuing theory involves an analysis of arrivals at a controlled facility, such as a modular construction plant, and an analysis of the processes currently in place to modularise them (Green & Armand, 2010). The result is a set of conclusions that aim to identify flaws in the modular construction system and suggest how they can be ameliorated.

The origin of queuing theory can be traced to the early 1900s in a study of the Copenhagen telephone exchange by Agner Krarup Erlang, a Danish engineer, statistician, and mathematician (Mallieswari, 2020). The study led to the Erlang theory of efficient networks and the field of telephone network analysis. Using queuing theory, the study of a line would break it down into six elements: the arrival process, the service and departure process, the number of servers available, the queuing discipline (such as first-in, first-out), the queue capacity, and the numbers being served. Creating a model of the entire process from start to finish allows the cause or causes of congestion to be identified and addressed. Furthermore, queuing theory is essentially a vehicle for cost analysis (Singer & Donoso, 2008). It would be prohibitively expensive, or indicative of not having very many customers, for most businesses to operate in a manner so that none of their customers or clients ever had to wait in line. Therefore, construction modular businesses use information gleaned from queuing theory to set up their operational functions to balance the cost of modular construction and the inconvenience caused by the sequential procedure of adopting modularity methodology.

Furthermore, queuing theory can be used to analyse how resources are allocated in modular construction projects (Wang et al., 2008). For example, understanding the flow of workers and equipment makes identifying bottlenecks and reducing resource waiting times possible, leading to more efficient modular construction processes. Queuing theory can also help in scheduling workers on a modular construction site by predicting the waiting time and the utilisation of available resources, such as equipment or material delivery, and finding the optimal sequence of tasks to minimise the overall waiting time (Seshaiah & Thiagaraj, 2011). In modular construction, the delivery and handling of material can be challenging; however, queuing theory can help optimise material delivery and handling on construction sites by analysing the movement of materials and predicting the waiting time at each stage. This can help ensure that materials are delivered to the correct location at the right time, reducing the overall construction time. Furthermore, the queuing theory can analyse the time for quality control inspections during modular construction (Cochran & Broyles, 2010). By optimising the inspection process, it is possible to reduce waiting times and improve the overall quality of the construction project. Broadly, queuing theory provides a valuable tool for improving the efficiency of construction processes, reducing waiting times, and increasing productivity. Construction managers can find bottlenecks and optimise resource allocation to improve outcomes by monitoring the movement of labourers, tools, and supplies.

3.2.4 Control Theory (2016)

Control theory is a field of control engineering and applied mathematics that deals with controlling dynamical systems in engineered processes and machines (Azimi et al., 2012). The objective is to develop a model or algorithm governing the application of system in-puts to drive the system to a desired state while minimising any delay, overshoot, or steady-state error and ensuring control stability, often to achieve a degree of optimality. Control theory dates back to the nineteenth century when the theoretical basis for the operation of governors was first described by James Clerk Maxwell (Afifi et al., 2020). Although control theory is significant in control systems engineering, which deals with designing process control systems for industry, other applications range far beyond this. Also known as the general theory of feedback systems, control theory is practical wherever feedback occurs. Thus, control theory deals with the behaviour of dynamical systems and the use of feedback to regulate modularity methodology outputs in the construction industry.

Modularity methodology in construction incorporates multiple dynamic systems, ranging from individual modules to assembly processes (Adekunle et al., 2023a, 2023b). The behaviour of these systems can be understood and managed using the framework that control theory offers. Construction managers can control the production and assembly processes by implementing control mechanisms. This allows for the smooth integration of each module into the overall structure and the creation of feedback loops for quality control. The feedback is crucial in modular construction to maintain high-quality standards (Bertram et al., 2019). Sensors and monitoring systems embedded in the production and assembly lines can provide real-time data on various parameters, such as dimensions, structural integrity, and environmental conditions. Control theory's feedback loops use this data to make continuous adjustments, ensuring that modules meet the required specifications before and during assembly.

Furthermore, control theory offers tools and methodologies to optimise and stabilise systems in modularity methodology for the construction industry (Goh & Goh, 2019). These principles can be applied to ensure that the various modules work together seamlessly, maintaining the stability and integrity of the overall structure; feedback loops allow for the quick identification and correction of performance deviations, keeping the system within desired operational parameters. Control systems in modular construction have been referred to as active control systems counteracting to the passive systems based on absorbing or dissipating energy, both classes of systems, plus combinations such as hybrid and semi-active systems, delimiting the field labelled as structural control (Wuni & Shen, 2020). Structural control originated in flexible space structures and was rapidly extended to civil engineering structures as a promising tool to protect bridges, buildings, and other structures against loads induced by earthquakes and wind, among others. It is always an incomplete task to reviewing the research effort in structural control, which nowadays involves a variety of disciplines and technologies such as mechanics, structures, materials, sensors, actuators, mathematics, and computation (Azadivar & Wang, 2000).

The ultimate goal of control theory is controlling systems, having those systems accomplish particular objectives or act in a desired manner (Leigh, 2004). Off-site construction shops usually involve repetitive operations, and project managers endeavour to maintain shop performance at a desired level, which makes control theory a potential option for achieving such a target. In control theory, the system or environment under consideration is called a plant (Ogata, 2002), and a controller is a mechanism that settles actions to implement in the plant to accomplish a specific objective (McKay, 2010). Control theory involves determining input variables of a plant that will cause the plant outputs to be as close as possible to a desired value, called a reference value. Its concepts' simplicity and application independence make this theory a favourable candidate for dealing with control-ling dynamic systems. However, the successful application of control theory in a system has several significant requirements. First, an objective regarding the system's future state should be defined. This objective is related to the system's performance over a certain period (Hasan et al., 2018). Generally speaking, the objective can be the desired performance of the system. Second, a set of possible scenarios regarding corrective actions needs to exist. Third, some methods for selecting the corrective actions to achieve the desired system performance must be developed. This requires a model capable of forecasting the effect of possible control actions on the system performance not only adequately and reliably but also comparatively quickly.

Ikuabe et al. (2025) presented a method to enhance project control by allowing timely corrective actions in the dynamic environment of off-site construction projects inspired by control theory concepts. The novelty of the proposed method, which is group method data handling (GMDH), lies in the new project control vision behind it, that is to say, in the case of performance deviation in the fast-paced environment of off-site construction, instead of reasoning, which can be a time-consuming process, the cause of deviation is not dealt with, and the ultimate goal is maintaining a desired performance promptly with minimum cost (Blank et al., 2020). The GMDH was utilised for the simple but effective and practical generation of a mathematical equation reliably representing the system under control. The method proposed to use this equation was to determine the best action for project deviation quickly. This not only helps off-site fabricators deliver projects on time, reduce overhead costs, maintain their reputation in the market, and avoid high-rate penalties but also gives them time to discover the reasons behind the deviation that occurred (Du et al., 2020). Challenges faced regarding implementing the proposed method in controlling off-site construction projects are also addressed, and solutions are provided.

3.2.5 Social Cognitive Theory (2019)

The social cognitive theory posits that humans' psychosocial functioning is governed by a triadic reciprocal causal interaction between cognitive, behavioural, and environmental or situational contexts (Ferdous et al., 2019). In applying this theory to modular construction methodology, it can be hypothesised that if the internal dispositions of project participants are understood and the situational context in which they operate is known, their behaviours can be predicted (Drira et al., 2007). This hypothesis, if valid, has significant implications for behaviour management in modular construction projects. Modularity methodology in construction involves distinct processes and collaboration across various stages, including design, prefabrication, and assembly. Understanding participants' cognitive and behavioural tendencies can enhance the management of several critical areas (Hayes et al., 2013). By predicting behaviours based on social cognitive principles, safety protocols can be tailored to address specific risks associated with modular construction, such as handling prefabricated components and on-site assembly. This proactive approach can lead to a safer working environment.

Social cognitive theory originated from social learning theory, which fundamentally was a theory of learning based on imitation or behavioural mimicry. According to Pajares (2002: 116), social cognitive theory (SCT) emerged out of the need to recognise the central role of "cognitive, vicarious, self-regulatory and self-reflecting processes in human adaptation and change". In this theory, human functioning derives from the causal relationship among psychological factors, behaviours and situational factors, which operate as interacting determinants that influence each other bi-directionally. These relationships do not suggest equality in the degree of influence that the three factors have on each other or that they influence each other simultaneously. That notwithstanding, it has been argued that the bi-directionality of influence means that situations are as much a function of a person or cognitive factors as a person's behaviour is a function of the situation, which, according to Nielsen (2017) is an indication that people self-regulate their behaviour in so far as they rely on cognitive supports and manage relevant environmental cues and consequences. This last point is quite relevant from an organisational perspective and, indeed, from a construction project perspective, where behaviour management is critical to delivering projects successfully for health and safety and other performance-related reasons. Various dimensions of behaviour affect project outcomes in construction projects. Giesekam et al. (2015) identified several vital behavioural changes fundamental to achieving the radical changes that construction clients have sought from the construction industry for decades.

Furthermore, Giesekam et al. (2015) opined that the behaviours associated with client orientation or what was described as a focus on the customer, which is particularly relevant at this time in the light of the recent Office of Fair Trade (OFT) reports into collusion within the UK construction industry. Client orientation encompasses the following dimensions, namely effort put into educating the client, the extent to which client satisfaction is monitored, the precedence of the client's needs, and the amount of contact and communication between the project organisation and the client. It is widely recognised that customer focus is a precursor to success. As a result, it has been argued that contractor organisations must be customer-focused in understanding and fulfilling client expectations. In Kutsch and Hall's (2005) study on the competencies of project managers, high-performing managers were found, among other tendencies, to show a marked propensity for seeking information about the real, underlying needs of clients beyond those overtly expressed within contractual documents and terms of appointment. The study also exhibited a responsible attitude towards dealing with customer service problems rapidly and efficiently. These all point to appropriate behaviour dimensions that relate to the client. Applying SCT to this context would imply, in the words of Kubjana et al. (2024a, 2024b), that the solutions or choices made by the construction project organisation (CPO) concerning each dimension and the effort put into pursuing them are reflected in the dynamic reciprocal relationships between members' perceptions and attitudes towards the operationalisation of organisational goals; members' day-to-day goal-directed behaviour; and the presence and quality of organisational systems and subsystems to support the goal-directed behaviour. This implies that assessing a particular dimension of behaviour also requires an assessment of members' perceptions/attitudes (personal factors) and the presence or quality of systems to support such behaviour (environment). In other words, a complete picture of behaviour can only be obtained when all three aspects have been investigated.

3.2.6 Technology Acceptance Theory (2019)

The technology acceptance theory is an information systems theory that models how users accept and use technology (Momani & Jamous, 2017). The actual system used is the endpoint where people use the technology. Behavioural intention is a factor that leads people to use technology. It is significant to note that user acceptance and confidence are crucial for further developing any new technology. Businesses have facilitated or planned to facilitate their business platform to increase personnel efficiency, marketing improvement, cost reduction, and profitability (Innella et al.,

2019). Therefore, recognising the needs and acceptance of individuals is the beginning stage of any business; this understanding would be helpful to find the way to future development.

Acceptance is "an antagonism to the term refusal and means the positive decision to use an innovation" (Taherdoost, 2021, p. 1). Decision makers need to know the issues that influence users' decisions to use a particular system so they can take them into account during the development phase. Several researchers developed theories and models to describe and analyse user acceptance, and each model determines different factors to explain user acceptance. The question about user acceptance concerns all researchers who want to predict which technologies will prove appropriate for an organisation. On the other hand, user acceptance is critical to successfully implementing any new technology (Taherdoost, 2018). Additionally, it is significant to note that technology's features play a vital role in determining whether individuals involved in an activity will use it or not. Thus, understanding the users' perception of adopting new technology could help facilitate further growth in implementing that technology.

Thus, academicians and practitioners are interested in realising the factors that drive users' acceptance or rejection of new technologies (Almaiah & Al Mulhem, 2018). Answering this question may help stakeholders formulate better methods for designing, evaluating and predicting the users' response to the new technologies. Technology acceptance theories have been applied in a wide variety of domains to understand and predict users' behaviour, such as voting, dieting, family planning, donating blood, women's occupational orientations, breast cancer examination, choice of transport mode, turnover, using birth control pills, education, consumer's purchase behaviours, and computer usage (Liang & Chao, 2008). Several research studies in the field of technology acceptance developed frameworks and models to assess the usage of new technologies; these models introduce factors that can affect user acceptance.

An exceptional technology and application might be designed and developed; however, the project will fail if people do not get involved and do not use it. Thus, user acceptance is an undeniable key to further implementing and developing any technology and application (Chan et al., 2011). In other words, factors influencing user acceptance should be emphasised to increase technology usage and user adoption. Recognition of the needs and acceptance of individuals is the beginning stage of any business; this understanding would help find a way for future development. Thus, academicians and practitioners are interested in identifying the factors driving users' acceptance or rejection of new technologies.

3.2.7 Stakeholder's Theory (2017)

Stakeholder theory addresses business ethics, morals, and values when managing stakeholders involved in a project or organisation (Harrison et al., 2015). It seeks to optimise relations with stakeholders, improving efficiencies throughout the project or organisation. Stakeholder theory is used in many fields, such as project management, corporate social responsibility, strategic management and business ethics. The first scholar to define stakeholder theory was organisational theorist Ian Mitroff in his book Stakeholders of the Organisational Mind, which was published in 1983. Shortly after that, an article about stakeholder theory was released in 1983 by philosopher and professor of business administration R. Edward Freeman in the California Management Review. Freeman does not cite Mitroff as a source. Instead, he attributes stakeholder theory to discussions at the Stanford Research Institute. He published his book Strategic Management: A Stakeholder Approach shortly after the article. That action is why stakeholder theory is sometimes also referred to as Freeman's stakeholder theory.

In Freeman's book, he identifies and models stakeholder groups within a corporation, describing and recommending ways to manage their interests and determine who counts from the company's perspective (Freeman & Velamuri, 2023). He added that increasing value for stakeholders will improve the construction business in all aspects. Stakeholder theory notes that there are several interested parties or stakeholder groups that must be included under the umbrella of stakeholders, such as the company's employees, customers, suppliers, financiers, communities, governmental bodies, political groups, trade associations, trade unions and even competitors, as they too can impact the company. One of the values identified by stakeholder theory includes greater productivity across the organisation (Harrison & Wicks, 2013). If employees, who are considered stakeholders, feel as if they are being valued, then they will work harder and be more productive. This also means that companies will have excellent retention of not only their employees but also their clients. If productivity increases, the product or service delivered to the clients will improve. With that improvement comes more customer loyalty, especially as they are a company's many stakeholders when making decisions. Customers are also more likely to refer other customers to the company. This leads to more financier investment (Qi et al., 2021). They too, of course, are stakeholders. While sometimes they are considered the only stakeholders or the most important to a company as they hold their hands on the capital level, they are connected to other stakeholders. As other stakeholders are valued, the company's value grows, and investors are more likely to add money to production to take advantage of this increased market share.

Brower and Mahajan (2013) opined that stakeholder theory drives more than profits and productivity. There are ethical benefits to practising it as well. Construction companies find that the workforce's mental health improves as job satisfaction increases. It will also elevate the company's socioeconomic status in the local community. When one company practises stakeholder theory, it creates healthy competition among other companies, whereby all can thrive and help benefit their stakeholders. Stakeholder theory is a component of more extensive stakeholder management, which creates positive relationships with stakeholders by managing their expectations and objectives. In order to control this process, a strategic plan is required. To begin, stakeholders are identified, their influence and interest determined, and a communication plan is devised to keep them informed. However, not all stakeholders are created equally. This does not mean some are more important than others; it is just that prioritising offers a structure for managing them effectively. Again, to manage stakeholders and develop an effective strategic plan, it is crucial to understand the stakeholders (Yang et al., 2023), knowing their financial or emotional interests in the outcome of the work, what motivates them, what data is required from them, how they want to receive information, what they think of the job you are doing, and who influences their opinions.

3.2.8 Information and Communication Technology Theory (2022)

The construction sector is considered one of the most significant investment-draining sectors worldwide, with a strong influence and contribution to economic development (Ikuabe et al., 2025). However, this sector faces considerable barriers and challenges, such as low performance, frequent construction delays, quality issues and cost overruns. Therefore, finding solutions to improve construction project performance and make project management knowledge more efficient is the main challenge for organisations in achieving client requirements (Adekunle et al., 2022a, 2022b). One of these solutions is to apply new technologies and implement adequate information and communications technology (ICT) theory in construction projects.

ICT theory can be described as using technologies involved in the collection, transport, retrieval, storage, access, presentation, and transformation of all forms of information (Zhong et al., 2024). ICT applies to information and communication technology tools, including software and computer hardware. On the other hand, ICT infrastructure can be defined as the necessary components to manage and operate enterprise IT environments. An organisation can deploy its IT infrastructure in a cloud computing environment or computer facilities (Kubjana et al., 2024a, 2024b). These components include hardware, software, and networking and operating systems (OS), all used to provide IT services and solutions.

ICT utilisation is crucial in supporting project operations, and the organisational ICT infrastructure supports the performance of ICT implementation from the project level to the organisation level (Adekunle et al., 2022a, 2022b). Moreover, many construction companies have turned to ICT utilisation to respond rapidly to environmental changes and owners' demands. Creating new and innovative processes for projects is one significant way ICT can add value to an organisation's performance. In addition, if an organisation has a robust ICT infrastructure, IT can enhance coordination, communication, cooperation, and collaboration among team members (Qi et al., 2021). Although some research has suggested a positive correlation between ICT implementation and organisational performance, little empirical research has concentrated on the alignment between the ICT organisational infrastructure and ICT utilisation. These studies focus on the impact of IT infrastructure on projects and capital facilities or infrastructure performance. There are limited studies on the adoption of ICT by construction organisations and fewer studies regarding the effects of solid alignment of the ICT infrastructures and ICT utilisation on organisation performance (Eliwa et al., 2022). This is considering the relatively long period that computer integration has been advocated in construction literature.

Based on the available information, there is a lack of evidence about the business value of ICT utilisation in the construction industry (Aigbavboa et al., 2024). This scarcity of evidence could be why ICT usage, especially for project processes, seems limited. Therefore, it is essential to know how this ICT usage impacts project team–owner relationships, general organisation outcomes, and the success of construction projects. However, ITC utilisation alone is not the perfect solution with a utilisation strategy. Accordingly, the alignment of ICT infrastructure is crucial to the overall organisation's performance. Many studies have been conducted on ICT adoption in the construction industry and its relation to construction project performance. These studies generally cover ICT utilisation and factors influencing its implementation as part of the construction management process. It also discusses the expected benefits of using ICT in project management and construction. Considering the type of ICT usage, Pour Rahimian et al. (2020) examined the extent to which ICT, in general, is being used in the execution of construction industry projects. On the other hand, Elghaish and Abrishami (2020) investigated the potential of a specific ICT, namely the 4D BIM technology, as a productive tool in project management. Oztemel and Gursev (2020) provided a valuation model that measures the utility of electronic networking technologies in construction project activities' progress as a type of ICT. The model highlights how such technologies enhance communication, coordination, and information sharing among project stakeholders, thereby reducing delays and errors. This also emphasises the role of ICT in improving decision-making efficiency and overall project performance through real-time data accessibility.

3.2.9 Sustainability Theory (2022)

In its literal rudiments, sustainability means a capacity to maintain some entity, outcome, or process over time (Lim, 2022). Agriculture, forest management, or financial investment might be deemed sustainable, meaning the activity does not exhaust the material resources on which it depends. An analogous use of the term 'sustainability' refers to dependent social conditions; for example, a peace treaty, an economic policy, or a cultural practice may be called sustainable if it does not exhaust a political community's support (Gal & Gan, 2020). In its increasingly common use, sustainability frames how environmental problems jeopardise the conditions of healthy economic, ecological, and social systems. On a global scale, the political challenge of sustainability raises a set of fundamental problems and comprehensive goals (Steg & Vleck, 2009). By focusing on the ecological dependency of economic and social systems, sustainability reflects the mutual effects of environmental degradation caused by human activities and the perils to human systems presented by global environmental problems. The concept of sustainability thus raises a starkly fundamental question regarding human activity successfully maintaining itself and its goals without exhausting the resources on which it depends.

Asking such questions regarding human activity directs attention toward the planetary impact of human activity and its durability over time (Weissmann, 2020). It therefore provokes reflection on the manner and purposes of global human society. Problems such as biodiversity loss and climate change point to the global reach of humanity's powers and the scale of its risk. Mitigating their impact and risk seems to require reform across many human systems, namely financial, political, production, energy, transportation, communication and education. However, those reforms could complicate other goals of the international community, such as overcoming extreme poverty and protecting human rights (Chopra et al., 2021). The mutual relationship is that sustainability brings into view and illuminates the dependency of those goals on ecological conditions; overcoming poverty cannot, in the long run, stand in competition with protecting sufficient biodiversity. The practical challenge of sustainability is to find specific ways to pursue those distinct goals that conform to their mutual relation (Gupta et al., 2021). Therein lies the possibility of maintaining (or developing) a decent quality of human life over time. Thus, sustainability concepts confront societies with new moral objectives: resources to be sustained, resources that the dramatic expansion of human systems may imperil, resources to be protected, goals that must be pursued, and the shared foundation for pursuing the goals.

Within construction institutions, the establishment's purpose can contextualise those practical objectives and their relation to ecological and social systems (Phulwani et al., 2020). For a university, sustainability may primarily be an aspect of how it manages its energy and food systems in relation not only to its budget but also to its sense of civic leadership and educational mission. For a corporation, sustainability may mean anticipating how the reflexivity of ecological, economic, and social systems will determine market conditions over more extended periods than those covered by quarterly or annual reports (Hogg & Shah, 2010). At local and global levels, sustainability directs practical attention to the complex mutuality of human and ecological systems. Economic health, ecological integrity, social justice, and responsibility to the future must be integrated to address multiple global problems within a coherent, durable, and moral social vision. That inclusive scope and prospective vision make sustainability ideologically absorptive and politically popular. Sustainability is used to argue for and against climate treaties, for and against free markets, for and against social spending, and for and against environmental preservation (Farioli et al., 2018). Some critics have dismissed sustainability as either conceptually meaningless or too broad and open to competing ideas to be politically useful. However, as long as the disagreements generally recognise mutual feedback between human and ecological systems, they reflect substantive differences about what to sustain over time. Therefore, sustainability produces a significant discursive arena for a new moral and political debate (Prothero et al., 2011). Because those considerations are urgent and essential, diversity of opinion and conceptual disagreement should be expected.

3.2.10 Resource Dependence Theory (2009)

Hillman et al. (2009) devised the resource dependence theory to explain how organisations' behaviour is affected by the external resources they possess. They propose that firms change and negotiate with their external environment to secure access to the resources they need to survive. This means that a firm's competitiveness is determined by how it deals with its external resources. Damanpour et al. (2018) believe that this is, in fact, more important than their internal resources. Resource dependency theory is one of the two resource-based perspectives in strategic management, contributing significantly to its literature. It explains the impact of the environment on a firm in terms of resources and how firms can use strategies to manage this impact (Jiang et al., 2023). This perspective has always looked at organisations or firms as the unit of analysis, maintained an external perspective, and therefore tried to fathom the inter-organisational actions of a firm in terms of mergers, acquisitions, vertical integration, board interlocks and succession planning.

Resource dependency theory examines the relationship between organisations and the resources they need to operate (Tan et al., 2017). Resources can take many shapes or forms, including raw materials, workers, and funding. If one company maintains most of a resource, another company will depend on it to operate, creating a symbiotic relationship. Too much dependency creates uncertainty, which leaves organisations subject to the risk of external control. The government or other organisations may impose external control and can significantly affect operations, such as funding or personnel policies (Hossain et al., 2022). Managers strategise alternative business plans in order to lower this risk. Resource dependence theory allows construction establishment leaders to understand their competition better, be aware of their dependence on a particular company or resource and see how diverse their business practices are. This motivates the adoption of modular construction that aids faster delivery of construction projects to clients.

The elementary proposition of resource dependence theory is the need for environmental associations between the firm and external resources (Du & Li, 2020). In this viewpoint, directors link the firm with outside factors by co-opting the resources needed to survive. This means that the board of directors is a significant mechanism for absorbing critical elements of environmental uncertainty into the firm. In other words, the theory examines the relationship between organisations and the resources they need to operate (Tan et al., 2017). Resources come in many forms, such as raw materials, funding, and employees. Getting these external resources is essential to any business's strategic and planned management.

Resource dependence theory supports the idea that resources are vital to organisational success and that access and control over resources are a basis of power (Mao et al., 2021). Resources are often organised by organisations that are not in the control of the organisation demanding them, implying that strategies must be carefully measured to maintain open access to resources. When a corporation owns all of the stock of a particular external resource, it will make other companies reliant on them to operate (Sarmidi et al., 2013). This creates a reciprocal relationship. Too much reliance creates uncertainty, weakness, and the risk of external checks being levied on the organisation. External checks can be enforced by governments and other organisations, which can substantially affect a business's operational activities, for example, when it comes to financing or hiring.

3.3 Modularity Methodology Models

Modularity methodologies have become increasingly popular in engineering and product design. These methodologies involve breaking down a complex system or product into smaller, more manageable components or modules. The modular approach provides numerous benefits, such as improved maintainability, reusability, and scalability. This study delves into the conceptual model of modularity methodology, exploring how it works, the different types of modularity, and its applications. This analysis provides a comprehensive understanding of modularity methodologies and their potential for enhancing the efficiency and effectiveness of complex systems and products.

3.3.1 Project-to-Product Model

The project-to-product model (PPM) was developed by Bertram et al. (2019) on behalf of McKinsey Global Institute (MGI). The model explains the more significant long-term opportunity of shifting to a modular construction model based on more efficient manufacturing-style production systems and prefabricated components. While this has been tried in various guises, it has never achieved full scale nor demonstrated the revolutionary productivity gains of which it should be capable. An increasing number of developers are intrigued by modular construction's potential but unsure how to pursue it in a way that guarantees reliable advantages (Xie, 2015). A good starting point for developers is identifying the segments of a portfolio where volume, repeatability, and retained ownership come into play. These can be designed as a 'product core' consistent across developments. These products may need to be tailored for a modular approach (for instance, reducing basements and bespoke ground floor designs, changing room widths to maximum road transport limits, or minimising variability). Using prefabrication while offering a degree of customisation, such as enabling clients to choose interior finishes and altering the façade and layout, will be crucial to satisfying end customers and local authorities (Velamati, 2012). The model inferred that developers should understand and optimise the strategic trade-offs in the products they commission and develop between quality, cost, and time savings.

Delivering projects in a new way challenges the traditional role of engineering and construction firms (Peltokorpi et al., 2018). While modules must still be assembled, on-site construction may become a more minor and commoditised part of the value chain. Today, general contractors manage complex projects with many subcontracted trades and shoulder execution risks. However, they risk being cut out of a value chain focused on simple module assembly with high cost and schedule certainty. Traditional engineering and construction firms can counter this risk through modular construction by moving into development, consulting, and planning (Smith & Rice, 2017). They can also aggressively make use of modules to gain margin advantages over competitors for what looks to be an extended transitionary period. Some may team up with module manufacturers. Finally, they can focus on highly complex projects that demand more on-site work since custom projects will not disappear, even in a more modular world (Smith & Rice, 2015). Even for firms that maintain a traditional focus, accelerating digitisation, being open to new collaborations, and keeping operations lean will be critical to competing in the future.

The project-to-product model thus emphasises that the variables affecting modularity methodology are supply chain, productivity, digitalisation, regulatory considerations, sustainability and design variables. Bertram et al. (2019) concluded that growing evidence of disruption in the construction industry has been discussed for decades. This disruption will result in winners and losers, and the construction landscape will look very different owing to the introduction of modular construction, making it an exciting sector for intelligent investors seeking alpha now. Investors should seek to understand the markets that will most likely be disrupted and the detailed trends, strategies, and capabilities that will set the winners and losers apart.

3.3.2 Liew & Chan Model

Modular construction shows excellent potential to improve construction productivity and efficiency. It is believed that modular construction will shape the construction industry's future. The Liew & Chan model (2019) summarised the key points to implementing modular construction methods for high-rise buildings, as shown in Fig. 3.1. These points include construction productivity, efficiency and safety, sustainability, transportation and installation. As the transportation and lifting requirements constrain the weight and size of a module, a composite lightweight modular system incorporating the advantages of both concrete and steel is proposed (Chen et al., 2017). Composite beams and concrete-filled tubular columns are proposed by the model for modular design, inheriting the merits of the concrete system of being durable, fire-resistant, water tightness and soundproof and carrying the advantages of the steel system of being flexible and long span, lightweight, and fast assembly. Lightweight concrete slabs and lightweight partition walls are proposed to reduce the weight of a module by up to 50%, leading to lower crane capacity needed.

Using composite columns in modular design helps keep the column size the same throughout the entire building, particularly for high-rise buildings with module repetition and section size standardisation. The modular unit has a double beam system in which the ceiling beam supports mechanical, electrical and plumping (MEP) services while the floor beam supports the floor loads. This results in less

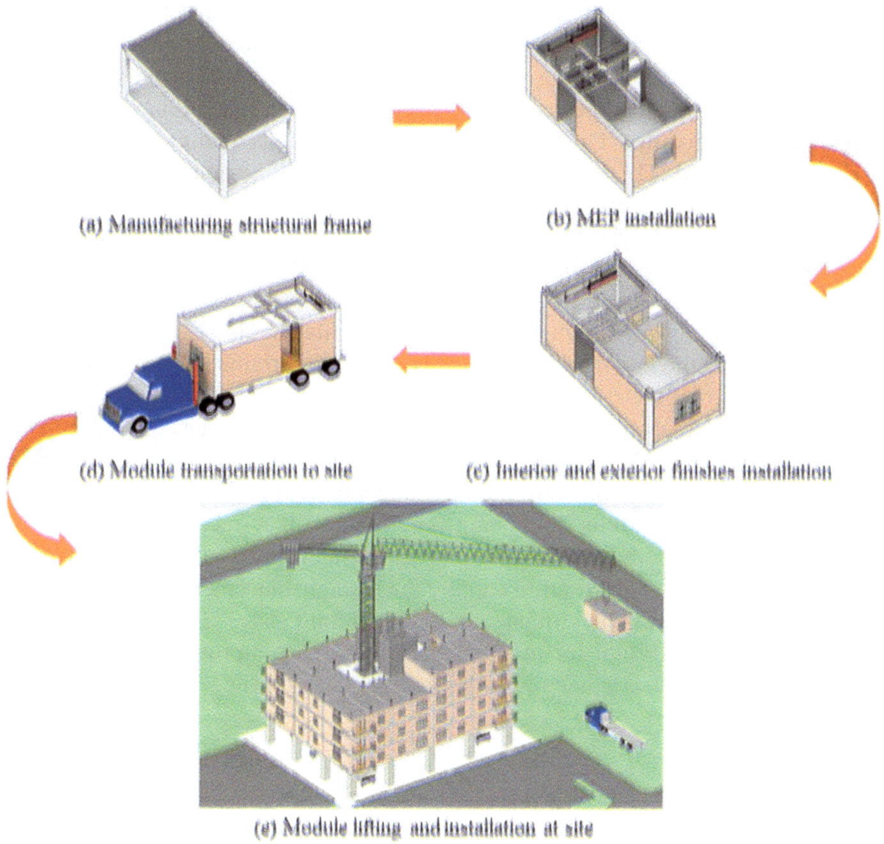

Fig. 3.1 Liew & Chan's model. *Source* Liew and Chan (2019)

available headroom in modular buildings, especially when a large beam span is required. Therefore, composite beam design coupled with a slim floor system is proposed to reduce structural height and increase the available headroom by up to 25% compared to conventional modular steel design. A long-span modular design of up to 12 m is proposed to reduce the number of columns by up to 60%, allowing for more open space and fewer connections to be installed (Li et al., 2017). With this, modular construction can be extended from residential and commercial buildings to more comprehensive applications, such as industrial buildings that require larger open space but higher design floor loads.

However, increased structural depth due to the longer span encourages the novel use of a slim floor beam system to reduce the module height (Zhang et al., 2016). The preliminary studies show that the slim floor beam system is efficient for module spans from 9 to 12 m in steel usage compared to conventional composite beams. A lightweight concrete slim floor slab can reduce the module's weight. With the increased number of connections in modular buildings, the design of fast and easy joining techniques is essential to improve the speed of construction. The joints must be robust enough to prevent progressive collapse due to column loss scenarios (Ramaji & Memari, 2013). As the modular building consists of many individual units, it is proposed that the modules can be connected via beams and columns from both upper and lower modules. The beam-to-column connections within a module should be designed as rigid. The joints between modules should be connected outside the modules for fully furnished modules. Novel joint designs that connect the modules via vertical rods and horizontal tie plates are proposed to facilitate such work. It is recommended that the vertical tie rod connection between the upper and lower module should be modelled as pin-connected. Conversely, the horizontal tie plate and the shear key connecting the modules horizontally should be modelled as fixed ended to simulate the fact that the tie plate is rigidly constrained at both ends by the clamping force from the columns (Pang et al., 2016).

Since the modules are connected at the corners, the floor slab of each module acts as a discrete floor diaphragm to transfer the horizontal load to the lateral force-resisting system as long as the inter-module connections are robust enough to transfer the horizontal loads without significant deformation. It is recommended that the global modelling of modular high-rise buildings should consider the semi-rigid joints and discrete floor slab behaviour to capture the load–displacement behaviour and to ensure global stability at the ultimate and accident limit states. This model provides ideas for composite design for modular construction to improve productivity and safety and reduce costs, human resources, and wastage on site.

3.3.3 Modularity Phases Model

The idea of the modularity phases model was developed by Gosling et al. (2016a, 2016b) on the premise that modular designs and approaches are a valuable means for managing complexity and helping rapidly respond to changing client requirements. Ethiraj and Levinthal (2004) opined that modularity involves breaking up a system into discrete chunks, which communicate with each other through standardised interfaces, rules, and specifications. Modularisation is a broad concept with various interpretations and meanings across research disciplines and market sectors. Gosling et al. (2016a, 2016b) argued that critical challenges for a modular system are finding the appropriate number and types of modules and defining their interactions and interfaces. This module addresses the problem of identifying and categorising modules in building projects.

The model combines different elements to develop a practical guide for practitioners. It is organised using general project phases with recommendations developed for practitioners that link to different parts of the model. Firstly, a grid layout supports the planning and design phase. This supports repeatability and standardisation in the design stage and later in the project phases. Secondly, a formalised product architecture should be developed to support design choices. This advice builds on general design guidelines by Ulrich (1995) but also extends the product hierarchy of a building system proposed by Schoenwitz et al. (2012). It was third, linking the product architecture to the four strategies, which include positioning the building, assembly of volumetric modules, assembly of non-volumetric subassemblies and then manufacture and assembly of materials. The strategies are employed in the preconstruction phase to establish the level at which they will be used. This helps establish a physical delivery strategy for the modules. Fourthly, the strategy can be operationalised through the planning matrix in the preconstruction stage, where the costs and benefits of different approaches can be considered and analysed. These recommendations also require the evaluation of inputs, drivers, and constraints. This helps in the formulation of an appropriate strategy for the project.

In further explanation of the model, Gosling et al. (2016a, 2016b) also highlighted the need to think across project phases, including the integration of design, purchasing, and site operations. The analysis of perceptions of modularity at the different levels lends support to the distinction between the design theory view and an operations management view of modularity and the importance of uniting these different perspectives. This also supports the findings of Pan et al. (2012), highlighting the need for collaborative working between designers and site managers early in the project lifecycle, along with feedback from manufacturers to give insight into off-site possibilities, as offering the best opportunity for the benefits of modularisation to be realised. Finally, a systems thinking mindset is encouraged, based on the principles outlined in Gosling et al. (2013) in the approach to modular construction since it is essential to consider the whole and the role of modules and actors within it.

3.3.4 Hammad & Akbannezhad Model

The Hammad & Akbannezhad (2017) model depicts the overall framework, which comprises four principle modules: the first set of modules incorporates the input data required for the analysis. The second module is a building information modelling (BIM) representation of the project to be constructed, aggregating the input data. Associated with the BIM is a comparative module specifying the material composition required for each construction method adopted (Chen et al., 2010). The last module is a criteria analysis module that embeds the conditions necessary to compare the construction methods. Hammad and Akbannerzhad (2017) noted that data is required to assess the available construction methods. Modular construction relies on an environment that is drastically different from that of conventional construction. As a result, differences would arise in the project information regarding site preparation and building layout, material to be used, equipment deployed during the construction process, carbon factors of the associated material with each construction method, and the cost element related to the building components (Mao et al., 2013). Each information mentioned above forms a separate database, which the constructed BIM utilises.

Furthermore, a BIM is associated with a modular construction method to link the various information from the input database (Monahan & Powell, 2011). Depending on the size of the project and its use type, different methods will need different material compositions; hence, the building model produced through the modular methodology can vary. Modular construction significantly emphasises using timber and light gauge steel framing instead of brick and mortar in conventional construction (Ezcan et al., 2013). As a result, the BIM produced for each modularity method will have to reflect this difference in element composition. Materials and equipment can be mapped according to the modularity method used, and information for the latter databases can then be associated with the BIM of each modularity method.

The comparative module is where the BIM for modular construction is used to extract the necessary data required to evaluate the multiple criteria defined. The criteria are evaluated in a way where economic and environmental factors influenced by the construction methods can be assessed (Faghirinejadfard et al., 2015). Three criteria are incorporated within the comparative module of the framework: embodied carbon, construction project duration, and construction cost. The authors revealed that modular construction was more effective in lower embodied carbon, higher productivity and lower construction costs. Future research will incorporate additional parameters and criteria to compare modular and conventional construction methods.

3.3.5 Modularisation Strategies Model

The modularisation strategy model was developed by Peltokorpi et al. (2017) by combining the concept of modularisation strategy with earlier research on the outcomes of different modular product architectures and production systems. A theoretical framework is developed to categorise modularisation strategies according to building investment objectives. These objectives can be identified in a building's life cycle according to the phase where they exert the most effects (Gosling et al., 2016a, 2016b). Activities in the design phase enable innovative design solutions. Low costs, high quality and a tight schedule require modifications in the construction phase and flexibility in use and maintenance demands that the use or the post-construction phase be considered. The model indicates that modularisation strategies vary in organisation, product architecture, production system, and platform assets used according to an investment's objectives. This is in contrast to previous research that emphasises that modularisation always aims to gain benefits in the construction phase related to improved cost efficiency, quality or schedule (O'Connor et al., 2014; Razkenari et al., 2020). The model highlighted that in addition to those objectives, modularisation could aim at innovative design solutions if project stakeholders' knowledge assets related to modularisation are used when designing repetitive spaces or structures and connected to a fabricate-to-order strategy. On the other hand, flexibility in the use and maintenance of a building is possible if module suppliers are given higher autonomy to utilise their standard interfaces and components as platform assets that enable a scale economy (MacDuffie, 2013). This does not apply only during the project but also in post-construction services. The model revealed that modularisation strategies that aim at improved performance during construction can be located between the extremes. They often require module suppliers' standard offsite production processes but not fully standardised components or interfaces. Therefore, the project organisation's role in design and delivery can be more significant than when aiming at flexibility in use and maintenance. Based on the model, three fundamental modularisation strategies can be differentiated.

Firstly, it is an innovative design solution that requires the project organisation, involving the project owner, designers and contractors, to play a remarkable role in defining which features are included in the innovative design solution. The project organisation focuses on designing a crucial and repetitive space or element in a building as a separate module, such as a bathroom in a hotel or a patient room in a hospital. The component supplier can be hired to prefabricate and deliver the module. However, it performs a minor role in designing the solution because the supplier's inherent incentive is to use more scalable and mature design solutions. Owing to the project organisation's scarce resources, the modules often comprise only a part of the building. They are connected to an integral building structure via project-specific slots or bus interfaces (Erikstad, 2009). Existing knowledge and relationships can be scalable platform assets in innovation work, especially among designers and contractors.

The second strategy is improved performance in the construction phase through modularisation, whether related to cost, quality or time. This requires adopting a prefabrication and preassembly-dominated modularisation strategy using an industrial production environment (Jonsson & Rudberg, 2013). Task division between the project organisation and component suppliers may vary according to the context and available resources. The primary platform assets used for an improved outcome are not necessarily standard components but are standardised processes to manufacture, preassemble, deliver and assemble components onsite.

Flexibility in use and maintenance, which requires that single components can be removed and replaced, is the third strategy. Volumetric preassembly is typically impossible as volumetric modules, such as bathrooms, often have complex and fixed interfaces to other structures. In order to enable replaceability during the building's life cycle, design rules about components and their interfaces should remain relatively stable (Papazoglou & Van Den Heuvel, 2006). These design rules can either be industry-wide standards or be managed by the module supplier to facilitate component-level replacements within the modular system. Thus, we argue that for flexibility in use and maintenance, there is a need for a supplier who owns the modular product system and can provide supplemental components and services during the building's use phase. Suppliers are often reluctant to adapt to new design rules; therefore, they should play a significant role in the project's design phase. Flexibility requires suppliers to utilise their standardised components and interfaces as platform assets (Andrews, 2015). Industrialised suppliers can have the resources to provide more complete modular buildings than project organisations, including wholly modular buildings where only the earthworks and foundations need customisation according to specific circumstances. Integrating design and supply chain management enhances modular construction projects' efficiency, quality, and longevity. A complete modular architecture enables the interfaces to be standardised across different functional components using a sectional interface architecture.

In summary, the model integrates the literature on modular product architecture (Salonitis et al. 2010), product platforms (Martin & Ishii, 2002; Schuh et al., 2010) and modularisation-related production strategies in construction (Choi et al., 2016; Gasper et al., 2012). The model's main contribution is expanding the current understanding of appropriate modularisation strategies by matching them with the building investment problems that modularisation can solve. The range of problems can vary from innovative design solutions to the traditional focus on project performance and flexibility in the building's use phase.

3.3.6 *Total Quality Management (TQM) Implementation Model*

Total quality management (TQM) is a management model based on the belief that an organisation can build long-term success by having all its members, from low-level workers to the highest-ranking executives, focus on improving quality and thus delivering client satisfaction (Al-Khalifa & Aspinwall, 2000). The model was developed by the father of TQM himself, Edwards Deming, who contributed to the revolution in Japanese manufacturing management, which led to the economic miracle of the 1970s and 1980s. In modularisation, the primary cause of the low yield rate of the placing module process was an accumulation of deviations due to workers' imperfectly placing modules on the lower floors. Another reason identified by the study of Phan et al. (2019) was the inadequate connection tolerance of the modular modules. These can be traced back to a poor organisational culture of quality. TQM is a model of management based on the principle of doing things right the first time (Shahin & Dabestani, 2011). It is widely seen as an approach to improving the quality of an organisation's goods and services through the involvement of all its members at every level.

In the context of the construction industry, TQM has been shown to reduce the occurrence of defective work. Alawag et al. (2023) examined the involvement of contracting firms in projects and the benefits derived from having a TQM-oriented culture, which include lower quality-related costs and higher employee job satisfaction. Although there is no one-size-fits-all recipe for a TQM implementation, it is observed that the most essential and crucial elements are to gain top management commitment, employee involvement, continuous improvement, training and customer focus. Linnenluecke and Griffiths (2010) provided essential steps for the parties in construction projects to inculcate a quality-oriented organisational culture. The TQM implementation model provides a path adapted from the earlier steps to implement TQM in modular construction. The parties involved are the design consultants, factory and construction managers.

Furthermore, as construction projects become more prominent, complex, and advanced, the importance of quality information for effective project management has increased. The accumulation of quality information can lower the defect rate in construction projects by increasing the efficiency of quality management. Reducing the defect rate positively affects cost and schedule in construction projects by decreasing the rework rate. In addition, with construction projects growing in size and complexity, the amount of quality information to be managed has also increased. Quality information is generated in all phases of construction projects, such as during contract change, design change, and nonconforming construction management, in addition to physical inspections for building components, such as construction and material inspections.

Meanwhile, modular construction has emerged to overcome the limitations of the existing site-oriented construction method (Gatheeshgar et al., 2021). An off-site construction (OSC) involves modularising a building into a panel- or volumetric-type unit. It combines the existing construction method of site-oriented operation with the production methods of the manufacturing industry. Specifically, modular construction involves transporting and installing building modules manufactured in a factory to the site (Molavi & Barral, 2016). The current site-oriented construction method exhibits low productivity; it is challenging to utilise production automation equipment owing to the narrow workspace. Further, because most construction activities are outdoors, site-oriented construction methods are greatly affected by weather conditions. In contrast, it is easy to use automated facilities in a modular construction factory, and the influence of weather conditions is not significant because the main processes are performed in indoor factories. Finally, the model postulates that modular construction can reduce construction time by simultaneously executing module manufacture in the factory and on-site construction processes, such as excavation and foundation construction. Therefore, modular construction can improve productivity by approximately 60% compared to the site-oriented construction method.

3.3.7 Risk in Modular Construction Model

Pervez et al. (2022) opined that the construction sector is crucial to a country's economy. One of these significant roles of the construction sector is to fulfil countries' infrastructure and housing needs. However, this traditional method poses various challenges, including on-site safety concerns, more extended project completion time, cost overruns, decreased quality, increased construction waste, inefficiency and environmental sustainability concerns. Many developed countries are now shifting their focus towards innovative construction techniques, such as modular construction, to resolve the issues mentioned earlier in conventional construction methods (Ingle & Waghmare, 2015). In this regard, off-site construction plays a significant role in resolving the current issues in conventional on-site construction. Modular construction is the complete off-site construction in which 80–85% of the work involved in the construction stage is carried out in a factory. In contrast, the remaining tasks, such as excavation, laying of foundation and module installation, are performed at the construction site.

The benefits associated with modular construction are workflow continuity, reduction in construction waste, reduction in project completion time, reduction in carbon emissions, reduction in several on-site contractors and improved workers' safety (Lu et al., 2018). Considering the increased adoption of modular construction owing to its well-documented benefits, precisely project completion time, cost and construction waste reduction, other less-developed countries can also use it as a sustainable alternative to conventional construction. However, adopting and implementing modular construction on a large scale is a complicated process associated with multiple risks and uncertainties. Modular construction is an innovative method for less developed countries with a unique scope, steps and interfaces. These unique characteristics are associated with various complexities that give rise to potential risks, posing novel challenges to decision-makers. That is why Pervez et al. (2022) developed a risk management model for modular construction.

Previous literature provides valuable evidence on risk factors in various construction projects (Pervez et al., 2021). Though some risk factors are mutual in several construction projects, other risk factors differ for every construction project owing to different requirements such as resources, technologies and management skills. Similarly, projects involving modular construction methods have specific risks. Modular construction requires a significant amount of initial investment, which has been highlighted in several studies (Kamali & Hewage, 2016). This initial investment incorporates various costs such as land, factory, equipment, machinery and labour. Owing to fragmented and interdependent segments in the supply chain, a high level of integration and coordination is required for successful project delivery.

The complexity of the supply chain in modular construction is subject to the risk of ineffective coordination among various project participants, leading to design defects, change orders, inefficiency and schedule delays. Stakeholders involved in modular construction projects are also from diverse backgrounds, each with value systems and objectives. Unlike traditional construction, where all stakeholders mutually transfer project risk to a client, project participants in modular construction are amalgamated into a single market, hence responsible for steady project delivery. Owing to the diversity of stakeholders, conflict of interest is a significant risk which may affect the quality and performance of modular projects.

Another significant risk is the limited expertise and experience in modular construction, highlighted in various research studies (Kamali & Hewage, 2017). Limited expertise can result in problems such as poor design, manufacturing, assembly, installation and erection practices. Moreover, lack of experience can also result in conflicts due to poor coordination between designers and manufacturers in the preliminary stages of construction, problems during production and schedule delays. Modular construction, a novel approach in various countries, lacks proper design codes and standards (Quale et al., 2012). This carries a severe risk since implementing a novel approach without regulatory guidance and policies may result in financial loss. Lack of government support is also significant since many countries have government-owned construction firms as clients. In the final explanation of the risk in the modular construction model, Pervez et al. (2022) concluded that to mitigate this risk, the prime focus should be improving the quality of manufacturing. Once a desired level of quality is attained, the next focus should be on financing these small and medium-sized firms to increase their production capacity. Modular construction firms can also collaborate with external partners that provide warehousing and storage services in cases where internal expansion is relatively less feasible from an economic point of view.

3.3.8 Smith Offsite Construction Model

The Smith offsite construction model was developed because an ideological and structural shift in the construction industry is occurring with the abandonment of traditional design-bid-build contracts toward design-build and integrated project delivery that rely on performance contracting and shared risk and reward. Modular construction is synonymous with integrated delivery. Research shows that projects that decide to utilise modularity after the design development phase have demonstrated a negative impact on both schedule and budget due to employing modular too late in the building delivery process. In explaining the model, Smith (2014) postulated that stakeholders on modular projects should ask specific questions across typical project phases. These questions speak to the off-site aid in meeting the cost, time, labour, site and programmatic goals for the project. The project design in integration with stakeholders for off-site manufacture, transport, assembly and disassembly if required. And then, the design of the project developed so that work is structured for what is done on-site and what is manufactured in the factory. The detailing developed in collaboration with the design team, general contractor, fabricator and installer; the design changes reduction and the orders placed in a short time frame to reduce cost. The fabrication performed with prototypes and lead times then reduced in coordination with the project team. Consequently, the site deliveries made just-in-time, loaded and delivered to minimise handling; and finally the assembly operations designed collaboratively and continuously followed to ensure safety, quality, time and cost parameters are met (Mostafa et al., 2016). Therefore, the model's construct includes "predesign, design, develop, detail, order, fabricate, deliver and assemble."

Furthermore, Smith identified some differences between modular project delivery and traditional on-site construction. First and foremost, there might be a requirement for a third-party agency or a state programme that regulates the design, approval and inspection of the building at the manufacturing plant and an on-site inspector for site works (Hong et al., 2016). The extent or impact of these programmes varies from country to country, as well as the associated costs. Secondly, transportation regulations are also handled county-by-country. Therefore, shipping from a country where a module is manufactured to another country where the site is located will require differing codes, permitting, lead cars, and associated fees. Each project should carefully ensure this is well considered. On average, modular manufacturers, for example, are approved to build within the country of location (Ramaji et al., 2017). Also, exceptional engineering may be required to ensure modules are structurally robust enough for the dynamic load associated with transportation. Often, manufacturers or dealers will have their engineers in-house, in addition to the engineer of record in permanent modular projects servicing a traditional design and build team contracted with the owner. The fourth consideration should be that pick points where a crane will lift the panel or module from the truck trailer and set it on site will be designed as part of the element structure. Wooden modules may use a wraparound belt strap. The most common cranes for setting elements are a truck-mounted hydraulic crane, crawler crane and, in exceptional circumstances, a tower crane (Yarlagadda et al., 2017).

A determination on the designated personnel to set the elements should be decided early during contract negotiations. Finally, the critical path method (CPM) is the most common method of construction scheduling where trades are stacked sequentially on site (Naqvi et al., 2014). In modular construction, concurrent scheduling overrides CPM, requiring a professional construction approach to construction scheduling, cost, labour, and supply chain management in a radically different way to the advantage of the project's overall schedule.

3.3.9 Tolerance Strategies Modularisation Model

Shahtaheri et al. (2017) opined that the function of a tolerance strategy modularisation model is to develop and analyse a series of design options from a total cost perspective by addressing both module production and project risk. Each design option has a tolerance value attributed to a critical dimensional and geometric variability factor. When developing the tolerance strategy, a critical variability factor is selected based on the designer's judgement. For a given strategy, the designer must also determine how many iterations to undertake (increasing the number of iterations will produce a more in-depth analysis [Lu & Korman, 2010]. However, this consequently becomes more time-consuming, meaning there is a balance between time invested and the quality of the tolerance strategy), and it must also ensure that each design iteration fulfils the required design criteria. The module production cost function is derived by considering materials, labour and equipment costs. In contrast, the risk function is quantified by considering costs for rework, project delays, safety, transportation claims, and other project factors (Shahtaheri et al., 2015). Then, an amalgamated or total cost function can be developed by adding the module production cost function and risk function together. Using the total cost function and associated tolerance values, an evaluation of design options can reveal which is optimal for managing risk related to geometric variability and tolerances.

Furthermore, Shahtaheri et al. (2017) postulated that the modular construction tolerance strategy model can be explained in four steps. The first step is selecting different design solutions or configurations for evaluation. This may involve making changes to the original design for the module or assembly or making minor changes to the design, such as changing the connection types, adding or removing bracing, or changing assembly component (member) sizes or types. Changing the design solution may involve re-assessing loading on the module due to transportation and handling at this step (Wong & Lau, 2015). The second step is "assessing the adequacy of the design solutions or configurations". Each alternative must be evaluated to ensure that it satisfies the design requirements for the project. This step will include a rigorous structural design check for a large structural module.

The third step is to develop the amalgamated or total cost function. The total cost function was defined previously as the module construction cost function plus the module risk function. The module construction cost function includes the labour, equipment and materials to produce the module (Taylor et al., 2009). The risk function represents the estimated costs associated with incurring risk. The risk function should include any risks affecting the module from the production phase to transportation, assembly, alignment, erection and connection to other modules or structures. The project risks must be characterised by estimating the probability of the risk occurring and the estimated impact (cost) of incurring the risk. The final step is to define the Pareto optimal boundary. Pareto optimal boundary, Pareto efficiency, or Pareto optimality represents a set of solutions superior to the other solutions (Zhai & Reed, 2014). Finding a better individual solution outside this front is impossible without making at least one worse selection. This means that the Pareto set solutions satisfy all the objectives and are better than all the remaining (dominated) solutions. The solutions belonging to the Pareto-optimal front, or the Pareto-optimal boundary, are known as the non-dominated set of solutions, while the remaining solutions are the dominated solutions. In this stage, the total cost of the module, including module construction and risk costs, are related to the corresponding geometric or dimensional variability measure, and the optimal boundary is identified.

3.3.10 Modular Construction Process Model

A systematic risk management approach is needed for modular construction so that stakeholders can consider project risks and plan accordingly to avoid budget overruns and schedule delays. This thought led to the formation of a modular construction process model by Li et al. (2013). In 2000, the Project Management Institute (PMI) introduced a uniform methodology for risk management in the Project Management Body of Knowledge (PMBOK). This methodology includes planning risk management, identifying risks, performing qualitative risk analysis, performing quantitative risk analysis, planning risk responses, and monitoring and controlling risks. The PMI methodology also designs a blueprint for project risk management, which will be utilised in formulating a modular construction process model. Meanwhile, the Construction Industry Institute (CII) has outlined a three-level risk analysis process: identification, deterministic, and probabilistic. In this risk analysis process, the model categorised modular construction into offsite construction, site preparation, module shipping, onsite construction, finishing, renovation and landscaping.

Deduced from the model's explanation is the fact that the industry has shifted from risk transfer to risk reduction. It also underscored that the lack of a joint risk management mechanism is a crucial barrier to the application of risk management. However, despite the many existing studies in risk management for conventional construction, a gap still exists in the scholarship about risk management for modular construction. Also, in addition to the many advantages of modular construction outlined in the literature, modular construction still involves risks, and these risks differ from those for conventional project delivery owing to the unique activities that modular construction involves. Generally, engineering is one of the distinct risks encountered in modular construction, as opined by Shahtaheri et al. (2017). Modular construction increases reliance on the use of machinery, automated design and drafting, and production lines since most tasks are completed in a factory setting.

Another distinct risk is occupational and cultural due to the nature of factory-based construction and the tasks involved. The demographic composition of the workforce varies from conventional construction in that it could include, to a greater extent, women, older adults and, in some cases, persons with physical disabilities. Socioeconomic factors are another distinct risk encountered in modular construction. This is due to the degree of acceptance among consumers of this type of construction. Finally, financial factors due to the construction process and payment schedules that differ from conventional construction introduce financial issues for builders. Modular construction often requires a significant upfront investment due to the need for early procurement of materials and manufacturing costs, which contrasts with the more phased payment schedules typical of conventional construction. This financial burden can pose challenges for builders, necessitating careful financial planning and management to ensure sufficient cash flow and project viability.

Furthermore, several factors vary in modular versus conventional construction (Kamali et al., 2019). For instance, weather conditions affect conventional construction more significantly than modular construction. This is because much of the modular construction process occurs in controlled factory environments, reducing delays and potential damage caused by adverse weather. These differences necessitate careful financial planning and management to ensure the viability and success of modular construction projects. In summary, while several studies have investigated risk management for conventional construction, there is a lack of scholarship examining the associated risks for modular construction.

3.3.11 Lean and Green Production Model

Lean production is a set of principles, methods and techniques to improve industrial production processes, maximise their value and reduce costs. Lean is a mainstream in the field of industrial engineering. In parallel to lean, pollution prevention and waste reduction to improve environmental performance have led to the development of the concept of green production for modular construction. Also, the green paradigm is a relevant issue in literature. Therefore, Caldarelli et al. (2022) postulated a model for the ever-increasing attention to modular construction's economic, social and environmental aspects. They suggested that the techniques introduced with lean and green production reduce business costs and improve sustainability. Lean concepts have recently been introduced in the construction sector, using lean construction. The combination of lean and green (L&G) has been trendy in the industrial world in recent years. An L&G (Legal and General) Index is introduced to implement an effective strategy towards the industry 4.0 revolution. This is because the only way to survive is to implement lean manufacturing on one side and green manufacturing (or better sustainability) on the other.

The impact of lean performance on environmental performance is discussed. Lean production may have on the one hand, beneficial environmental impacts, for instance, in terms of waste reduction and increased efficiency. On the other hand, productivity and business growth goals can be critical for sustainability. For instance, supply chain integration with Kanban increases the frequency of deliveries and transportation pollution emissions. L&G strategies are promising in the construction supply chain (CSC), which is important in impacting GDP and the environment. Usually, construction modular projects are customised and involve a significant number of engineer-to-order (ETO) components. CSC is the integration of two main parts: the fabrication part (off-site) and the installation part (on-site). While the off-site part looks like any other manufacturing industry, with a controlled environment in factories, the on-site part is different, with high volatility that can cause delays and cost increments. Off-site activities are relevant in modular construction, while lean production, which originates in the automotive sector, can be very effective. The benefits of lean construction are oriented towards material flow improvement, the enhancement of productivity, the elimination of waste and the reduction of delays.

3.3.12 Modular Integrated Construction Adoption Model

Rethinking the idea of modular construction is necessary to solve the difficulties the construction industry faces in complicated contexts such as the high-density high-rise scenario. Therefore, drawing on the theories of modularisation and production, the modular integrated construction adoption model is put forward and defined as a game-changing, disruptively innovative approach to transforming fragmented site-based construction of buildings and facilities into integrated value-driven production and assembly of prefinished modules with the opportunity to realise enhanced quality, productivity, safety and sustainability. For the simplification of terms, 'modular' is defined as the process of modularising fragmented site-based construction elements, components, and facilities to integrate prefinished modules to best suit value-driven production and assembly. It should not be understood as 3D volumetric modules per se.

In addition, 'integrated' is defined as integrating different construction activities and players in the vertical design and supply chains and integrating various functions and stakeholders in the lateral institutional settings. The integrated process should reflect the progressive approach to promoting modular integrated construction (MIC) in various construction sectors, such as hostels, hotels, and housing. It should be in an appropriate roadmap to address their strengths, weaknesses, opportunities, and challenges. The integrated process should also avoid regarding MIC as a technical approach per se but embrace it within the legislative, economic, social and technological contexts (Pan et al., 2012). MIC represents the highest end of offsite approaches with the most significant integration of production. MIC therefore advances the offsite constructed components and two-dimensional (2D) panellised systems such as precast concrete facades. MIC embraces the theories of modularisation, forces and relations of production, value engineering and lean construction and promotes learning from manufacturing industries.

The MIC adoption model addresses the features of high-density, high-rise building construction in Hong Kong (Liu et al., 2016). First, it helps to reduce reliance on on-site labour and the effect of site constraints on construction. Second, it enables cross-border integrated procurement to address the limited construction-related manufacturing capability. Thirdly, it enhances the life-cycle quality and accountability of buildings. The adoption model of MIC will lead to more opportunities to gain economic benefits from reducing site labour requirements and shortening construction duration, and also significantly enhance productivity and sustainability at both project and industry levels. Modularisation carried out by the MIC approach offers a golden opportunity to unleash productive forces, shape production relations, and enhance construction productivity, quality, safety, and sustainability. Hence, it is instrumental in transforming the construction industry.

3.4 Measuring Modularity Methodology

As earlier stated, modularity in construction refers to using standardised, interchangeable, and prefabricated building components to facilitate faster and more efficient construction processes. Measuring modularity in construction involves using specific methodologies to quantify the degree of modularity in building projects. Methodologies used to measure modularity in construction are modularity index, component count, unit count, assembly efficiency and product variety. This modularity index measures the degree of modularity in a building project by analysing the number and types of standardised building components used. The index is calculated by dividing the total number of standardised components by the total number of components in a building. A higher modularity index indicates a higher degree of modularity in the building design.

Regarding component count, modularity methodology measures the number of different building components used in a construction project. A lower number of component types suggests a higher degree of modularity. Unit count is the methodology that counts the number of standardised units used in a building project. Standardised units are pre-fabricated building components that can be easily assembled on site. A higher unit count indicates a higher degree of modularity in the construction process. On the other hand, assembly efficiency is the methodology that measures the efficiency of the assembly process by analysing the number of building components required to construct each unit. A lower number of components per unit suggests a higher degree of modularity in the construction process. Conclusively, product variety methodology measures the variety of building components used in a construction project. A lower product variety suggests a higher degree of modularity in construction. Measuring modularity in construction can help project managers and builders optimise the construction process, reduce costs, and improve the quality of building projects. Builders can pinpoint opportunities for development and make well-informed choices when selecting building components and construction procedures by employing these approaches.

3.5 Determinant of Modularity Methodology

Most empirical studies of modularity methodology have been based on one of the abovementioned conceptual frameworks. Modularity methodology plays a crucial role in the construction industry by enabling efficient and cost-effective project delivery. Several studies and authors have proposed different models of modularity methodology tailored explicitly for the construction industry. Design for manufacturing and assembly (DFMA), introduced by Liew and Chan (2019), focuses on designing construction components and systems, emphasising manufacturing and assembly considerations. The methodology promotes standardised, prefabricated elements that can be quickly produced off site and assembled on site. DFMA enhances construction efficiency, reduces material waste, and improves project quality by streamlining the manufacturing and assembly processes.

BIM, an integrated process for collaborative design, construction, and operation of buildings, has gained significant popularity in the construction industry, as opined by Hammad and Akbannezhad (2017). BIM enables creating and managing a digital representation of a building throughout its lifecycle. Using a centralised database, BIM makes it easier for many project stakeholders to collaborate, coordinate, and develop modularly. It makes it possible to integrate different systems, parts, and disciplines effectively, improving project visualisation, coordination, and collision detection. Prefabrication and modular construction methods involve manufacturing components or entire modules in a controlled factory environment before transporting and assembling them on-site. This approach, championed by authors such as Smith (2019), offers numerous benefits, such as reduced construction time, improved quality control, enhanced safety, and minimised disruption to the surrounding environment. Prefabrication and modular construction enable the repetition and standardisation of components, leading to increased efficiency and cost savings.

Lean construction, inspired by lean manufacturing principles, aims to minimise waste, maximise value, and improve project flow within the construction process.

Authors such as Innela et al. (2019) have contributed significantly to this methodology. Lean construction focuses on eliminating non-value-adding activities, optimising resource utilisation, and enhancing communication and collaboration among project participants. Lean construction lowers project delays, boosts productivity, and increases project predictability by putting modularity concepts such as component standardisation, just-in-time supply, and off-site fabrication into practice. According to Emmatty and Sarmah (2012), platform-based design and construction methodologies propose using standardised building platforms or typologies that can be adapted and repeated across multiple projects. These platforms consist of pre-designed modules, systems, and components that can be easily customised to suit specific project requirements. Platform-based approaches enable the efficient utilisation of knowledge, resources, and lessons learned from previous projects. They promote modularity, reduce design and construction time, and enhance overall project efficiency. Industrialised construction, advocated by authors, involves the application of industrial manufacturing principles and technologies to the construction process. This methodology emphasises using automation, robotics, and advanced manufacturing techniques to produce building components in a factory-controlled environment. Industrialised construction reduces dependency on labour-intensive traditional on-site procedures, increases safety, expedites building schedules, and improves quality control by utilising modularity.

3.6 Summary

This chapter looked at the theoretical underpinnings of the modularity methodology, highlighting important viewpoints and methods that influence how it is used in the building industry. It covered a number of models that offer organised frameworks for putting modular practices into practice with a focus on effectiveness, adaptability, and quality. In order to help practitioners evaluate performance, adaptability, and integration effectiveness, the chapter also examined ways to measure modularity methodology. Additionally, it identified key factors that affect successful adoption, such as market readiness, stakeholder collaboration, regulatory environments, and technological capabilities. Overall, the conversation made clear that the modularity methodology is a technical and strategic approach that necessitates careful theory, practice, and contextual alignment in order to produce the best possible outcomes.

References

Acebes, F., Pereda, M., Poza, D., Pajares, J., & Galán, J. M. (2015). Stochastic earned value analysis using Monte Carlo simulation and statistical learning techniques. *International Journal of Project Management, 33*(7), 1597–1609. https://doi.org/10.1016/j.ijproman.2015.06.012

Adekunle, P., Aigbavboa, C., Thwala, D., Oke, A., & Akinradewo, O. (2022a). Construction information management: The role of fourth industrial revolution tools. In *Human factors in architecture, sustainable urban planning and infrastructure* (Vol. 58, pp. 254–261). https://doi.org/10.54941/ahfe1002359

Adekunle, P., Aigabvboa, C., Thwala, W., Akinradewo, O., & Oke, A. (2022b). Challenges confronting construction information management. *Frontiers in Built Environment, 8*, 1075674. https://doi.org/10.3389/fbuil.2022.1075674

Adekunle, P., Aigbavboa, C., Otasowie, O., & Ikuabe, M. (2023a). Benefits of blockchain technology in the prefabricated construction industry. In S. M. Ahmed, S. Azhar, A. D. Saul, K. L. Mahaffy, & R. U. Farooqui (Eds.), *Proceedings of the 13th International Conference on Construction in the 21st Century, CITC 2023* (International Conference on Construction in the 21st Century; Vol. 2023-May). East Carolina University.

Adekunle, P., Aigbavboa, C., Otasowie, K., & Adekunle, S. (2023b). Benefits of robotic utilization in the prefabricated construction industry. In *Proceedings of the 31st Annual Conference of the International Group for Lean Construction (IGLC31)* (pp. 746–754). https://doi.org/10.24928/2023/0134

Afifi, M., Fotouh, A., Al-Hussein, M., & Abourizk, S. (2020). Integrated lean concepts and continuous/discrete-event simulation to examine productivity improvement in door assembly-line for residential buildings. *International Journal of Construction Management*, 1–12. https://doi.org/10.1080/15623599.2020.1793075

Aigbavboa, C., Adekunle, P., Ikuabe, M., & Otasowie, K. (2024). Digitisation of the construction industry for economic growth: Implementation strategies. In K. Arai (Ed.), *Proceedings of the Future Technologies Conference (FTC) 2024, Volume 4*. FTC 2024. Lecture Notes in Networks and Systems (Vol. 1157). Springer. https://doi.org/10.1007/978-3-031-73128-0_34

Akhavian, R., & Behzadan, A. H. (2014). Evaluation of queuing systems for knowledge-based simulation of construction processes. *Automation in Construction, 47*, 37–49. https://doi.org/10.1016/j.autcon.2014.07.007

Al-Khalifa, K. N., & Aspinwall, E. M. (2000). Using the competing values framework to investigate the culture of Qatar industries. *Total Quality Management, 11*(4–6), 417–428.

Almaiah, M. A., & Al Mulhem, A. (2018). Analysis of the essential factors affecting of intention to use of mobile learning applications: A comparison between universities adopters and non-adopters. *Education and Information Technologies, 24*(2), 1433–1468. https://doi.org/10.1007/s10639-018-9840-1

Andrews, D. (2015). The circular economy, design thinking and education for sustainability. *Local Economy*. https://doi.org/10.1177/0269094215578226

Azadivar, F., & Wang, J. (2000). Facility layout optimization using simulation and genetic algorithms. *International Journal of Production Research, 38*(17), 4369–4383. https://doi.org/10.1080/00207540050205154

Azimi, R., Lee, S., & AbouRizk, S. M. (2012). Applying basic control theory principles to project control: Case study of off-site construction shops. *Journal of Computing in Civil Engineering, 26*(6), 681–690.

Bertram, N., Fuchs, S., Mischke, J., Palter, R., Strube, G., & Woetzel, J. (2019). Modular construction: From projects to products. Capital Projects & Infrastructure, June 1–30. https://www.mckinsey.com/industries/capital-projects-and-infrastructure/our-insights/modular-construction-from-projects-to-products

Blank J., Deb, K., & Pymoo. (2020). Multi-objective optimization in python. *IEEE Access, 8*, 89497–89509. https://doi.org/10.1109/ACCESS.2020.2990567

Blismas, N., & Wakefield, R. (2009). Drivers, constraints and the future of offsite manufacture in Australia. *Construction Innovation, 9*, 72–83. https://doi.org/10.1108/14714170910931552

Brower, J., & Mahajan, V. (2013). Driven to be good: A stakeholder theory perspective on the drivers of corporate social performance. *Journal of Business Ethics, 117*, 313–331. https://doi.org/10.1007/s10551-012-1523-z

Campagnolo, D., & Camuffo, A. (2010). The concept of modularity in management studies: A literature review. *International Journal of Management Reviews, 12*(3), 259–283. https://doi.org/10.1111/j.1468-2370.2009.00260.x

Chan, F. K. Y., Thong, J. Y. L., Venkatesh, V., Brown, S. A., Hu, P. J. H., & Tam, K. Y. (2010). Modeling citizen satisfaction with mandatory adoption of an E-government technology (December 26, 2011). *Journal of the Association for Information Systems, 11*(10), 519–549 (2010). Available at SSRN: https://ssrn.com/abstract=1976951

Chen, Y., Okudan, G. E., & Riley, D. R. (2010). Sustainable performance criteria for construction method selection in concrete buildings. *Automation in Construction, 19*(2), 235–244.

Chen, Z., Liu, J., & Yu, Y. (2017). Experimental study on interior connections in modular steel buildings. *Engineering Structures, 147*, 625–638.

Choi, J. O., O'Connor, J. T., & Kim, T. W. (2016). Recipes for cost and schedule successes in industrial modular projects: Qualitative comparative analysis. *Journal of Construction Engineering and Management, 142*(10), 04016055. https://doi.org/10.1061/(ASCE)CO.1943-7862.0001171

Chopra, M., Saini, N., Kumar, S., Varma, A., Mangla, S. K., & Lim, W. M. (2021). Knowledge management for business sustainability: An extensive review using bibliometrics. *Journal of Cleaner Production, 328*, Article 129592.

Cochran, J. K., & Broyles, J. R. (2010). Developing nonlinear queuing regressions to increase emergency department patient safety: Approximating reneging with balking. *Computers & Industrial Engineering, 59*(3), 378–386.

Damanpour, F., Sanchez-Henriquez, F., & Chiu, H. H. (2018). Internal and external sources and the adoption of innovations in organizations. *British Journal of Management, 29*(4), 712–730. https://doi.org/10.1111/1467-8551.12296

Drira, A., Pierreval, H., & Hajri-Gabouj, S. (2007). Facility layout problems: A survey. *Annual Reviews in Control, 31*(2), 255–267. https://doi.org/10.1016/j.arcontrol.2007.04.001

Du, J., Jing, Q., Choo, K.-K.R., Sugumaran, V., & Castro-Lacouture, D. (2020). An ontology and multi-agent based decision support framework for prefabricated component supply chain. *Information Systems Frontiers, 22*, 1467–1485. https://doi.org/10.1007/s10796-019-09941-x

Du, W., & Li, M. (2020). Assessing the impact of environmental regulation on pollution abatement and collaborative emissions reduction: Micro-evidence from Chinese industrial enterprises. *Environmental Impact Assessment Review, 82*, Article 106382.

Eastman, C. M., & Sacks, R. (2008). Relative productivity in the AEC Industries in the United States for on-site and off-site activities. *Journal of Construction Engineering and Management, 134*, 517–526. https://doi.org/10.1061/(ASCE)0733-9364(2008)134:7(517)

Elghaish, F., & Abrishami, S. (2020) Developing a framework to revolutionise the 4D BIM process: IPD-based solution. *Construction Innovation: Information Process Management, 20*(3), 401–420. https://doi.org/10.1108/CI-11-2019-0127

Eliwa, H. K., Jelodar, M. B., & Poshdar, M. (2022). Information and communication technology (ICT) utilization and infrastructure alignment in construction organizations. *Buildings, 12*(3), 281. https://doi.org/10.3390/buildings12030281

Ellerman, D. (2018). Logical entropy: Introduction to classical and quantum logical information theory. *Entropy, 20*(9), 679. https://doi.org/10.3390/e20090679

Emmatty, F. J., & Sarmah, S. P. (2012). Modular product development through platform-based design. *International Journal of Product Development, 17*(3–4), 220–239.

Erikstad, S. O. (2009). Modularisation in shipbuilding and modular production. In *Innovation in Global Maritime Production.*

Ethiraj, S. K., & Levinthal, D. (2004). Modularity and innovation in complex systems. *Management Science, 50*(2), 159–173.

Ezcan, V., Isikdag, U., & Goulding, J. S. (2013). *BIM and off-site manufacturing: Recent research and opportunities*. UCLan—University of Central Lancashire, Brisbane, Australia.

Faghirinejadfard, A., et al. (2015). Economic comparison of industrialized building system and conventional construction system using building information modelling. *Jurnal Teknologi, 78*(1).

Farioli, F., Barile, S., Saviano, M., & Iandolo, F. (2018). Re-reading sustainability through the triple helix model in the frame of a systems perspective. In T. Marsden (Ed.), *The Sage handbook of nature* (pp. 10–37). Sage.

Ferdous, W., Bai, Y., Ngo, T. D., Manalo, A., & Mendis, P. (2019). New advancements, challenges and opportunities of multi-storey modular buildings—A state-of-the-art review. *Engineering Structures, 183*, 883–893. https://doi.org/10.1016/j.engstruct.2019.01.061

Freeman, R. E., & Velamuri, S. R. (2023). A new approach to CSR: Company stakeholder responsibility. In S. D. Dmytriyev, & R. E. Freeman (Eds.), *R. Edward Freeman's selected works on stakeholder theory and business ethics*. Issues in Business Ethics (Vol. 53). Springer. https://doi.org/10.1007/978-3-031-04564-6_12

Gal, A., & Gan, D. (2020). Transformative sustainability education in higher education: Activating environmental understanding and active citizenship among professional studies learners. *Journal of Transformative Education, 18*(4), 271–292.

Gaspar, H. M., Ross, F. M., Rhodes, D. H., & Erikstad, S. O. (2012). Handling complexity aspects in conceptual ship design. In *Proceedings of the 11th International Maritime Design Conference (IMDC 2012)* (pp. 339–354).

Gatheeshgar, P., Poologanathan, K., Gunalan, S., et al. (2021). Modular steel construction systems. *Structures, 33*, 101–114.

Gentile, P. D. (2013). Theory of modularity, a hypothesis. *Procedia Computer Science, 20*, 203–209. https://doi.org/10.1016/j.procs.2013.09.262

Giesekam, J., Barrett, J. R., & Taylor, P. (2015). Construction sector views on low carbon building materials. *Building Research & Information, 44*(4), 423–444. https://doi.org/10.1080/09613218.2016.1086872

Goh, M., & Goh, Y. M. (2019). Lean production theory-based simulation of modular construction processes. *Automation in Construction, 101*(December 2018), 227–244. https://doi.org/10.1016/j.autcon.2018.12.017

Gosling, J., Naim, M., & Towill, D. (2013). Identifying and categorizing the sources of uncertainty in construction supply chains. *Journal of Construction Engineering and Management*, 102–110. https://doi.org/10.1061/(ASCE)CO.1943-7862.0000574

Gosling, J., et al. (2016a). Defining and categorizing modules in building projects: An international perspective. *Journal of Construction Engineering and Management, 04016062*, 1–11.

Gosling, J., Pero, M., Schoenwitz, M., Towill, D., & Cigolini, R. (2016b). Defining and categorizing modules in building projects: An international perspective. *Journal of Construction Engineering and Management, 142*(11), 04016062.

Green, L. V., & Armand, G. (2010). Using queueing theory to alleviate emergency department overgrowing. In J. J. Cochran (Ed.), *Wiley encyclopedia of operations research and management science* (pp. 1–9).

Gupta, H., Kumar, A., & Wasan, P. (2021). Industry 4.0, cleaner production and circular economy: An integrative framework for evaluating ethical and sustainable business performance of manufacturing organizations. *Journal of Cleaner Production, 295*, 126253.

Haas, C. T., & Fagerlund, W. R. (2002). *Preliminary research on prefabrication, pre-assembly, modularization and off-site fabrication in construction.* University of Texas at Austin.

Hammad, A. W. A., & Akbarnezhad, A. (2017). Modular vs conventional construction: A multi-criteria framework approach. In *ISARC. Proceedings of the International Symposium on Automation and Robotics in Construction* (Vol. 34). IAARC Publications.

Harrison, J., Freeman, R. E., & Cavalcanti Sá de Abreu, M. (2015). Stakeholder theory as an ethical approach to effective management: applying the theory to multiple contexts. *Review of Business Management*, 858–869. https://doi.org/10.7819/rbgn.v17i55.2647

Harrison, J. S., & Wicks, A. C. (2013). Stakeholder theory, value, and firm performance. *Business Ethics Quarterly, 23*(1), 97–124. https://doi.org/10.5840/beq20132314

Hasan, A., Baroudi, B., Elmualim, A., & Rameezdeen, R. (2018). Factors affecting construction productivity: A 30 year systematic review. *Engineering Construction and Architectural Management, 25*(7), 916–937. https://doi.org/10.1108/ECAM-02-2017-0035

Hayes, S. C., Levin, M. E., Plumb-Vilardaga, J., Villatte, J. L., & Pistorello, J. (2013). Acceptance and commitment therapy and contextual behavioral science: Examining the progress of a distinctive model of behavioral and cognitive therapy. *Behavior Therapy, 44*(2), 180–198. https://doi.org/10.1016/j.beth.2009.08.002

Hillman, A. J., Withers, M. C., & Collins, B. J. (2009). Resource dependence theory: A review. *Journal of Management, 35*(6), 1404–1427. https://doi.org/10.1177/0149206309343469

Hogg, M., & Shah, H. (2010). *The impact of global learning on public attitudes and behaviours towards international development and sustainability*. Development Education Association.

Hong, J., Shen, Q., Peng, Y., Feng, Y., & Mao, C. (2016). Critical success factors for prefabricated construction projects. *Journal of Management in Engineering, 32*(6), 04016022.

Hossain, M. E., Islam, M. S., Bandyopadhyay, A., Awan, A., Hossain, M. R., & Rej, S. (2022). Mexico at the crossroads of natural resource dependence and COP26 pledge: Does technological innovation help? *Resources Policy, 77*, Article 102710.

Ikuabe, M., Aigbavboa, C., Adekunle, P., & Otasowie, K. (2025). Digital transformation in the construction industry: Mapping out the required leadership attributes. In GhaffarianHoseini, A., Ghaffarianhoseini, A., Rahimian, F., & Babu Purushothaman, M. (Eds.), *Proceedings of the International Conference on Smart and Sustainable Built Environment (SASBE 2024)*. Lecture Notes in Civil Engineering (Vol. 591). Springer. https://doi.org/10.1007/978-981-96-4051-5_157

Ingle, P., & Waghmare, A. (2015). Review of modular construction practices. *International Journal of Engineering Research*.

Innella, F., Arashpour, M., & Bai, Y. (2019). Lean methodologies and techniques for modular construction: Chronological and critical review. *Journal of Construction Engineering and Management, 145*(12). https://doi.org/10.1061/(asce)co.1943-7862.0001712

Jiang, H., Luo, Y., Xia, J., Hitt, M., & Shen, J. (2023). Resource dependence theory in international business: Progress and prospects. *Global Strategy Journal, 13*(1), 3–57. https://doi.org/10.1002/gsj.1467

Jonsson, H., & Rudberg, M. (2013). Classification of production systems for industrialized building: A production strategy perspective. *Construction Management and Economics, 32*(1–2), 53–69. https://doi.org/10.1080/01446193.2013.812226

Kalantarnia, M., Khan, F., & Hawboldt, K. (2010). Modelling of BP Texas City refinery accident using dynamic risk assessment approach. *Process Safety and Environment Protection, 88*(3), 191–199. https://doi.org/10.1016/j.psep.2010.01.004

Kamali, M., & Hewage, K. (2016). Life cycle performance of modular buildings. *Journal of Cleaner Production, 116*, 146–160.

Kamali, M., & Hewage, K. (2017). Development of performance criteria for sustainability evaluation of modular buildings. *Journal of Cleaner Production, 164*, 1240–1250.

Kamali, M., Hewage, K., & Sadiq, R. (2019). Conventional versus modular construction methods: A comparative cradle-to-gate life cycle assessment for residential buildings. *Energy and Buildings, 204*, 109479. https://doi.org/10.1016/j.enbuild.2019.109479

Khalfan, M. M. A., & Maqsood, T. (2014). Current state of off-site manufacturing in Australian and Chinese residential construction. *Journal of Construction Engineering, 2014*, 1–5. https://doi.org/10.1155/2014/164863

Kholopane, P., & Vandayar, C. (2014). Lean application in student finance department within a learning institution can lead to high academic throughput: A case study. In *Proceedings of PICMET '14 Conference: Portland International Center for Management of Engineering and Technology*, Infrastructure and Service Integration, Kanazawa, Japan (pp. 988–994).

Kubjana, L., Adekunle, P., Aigbavboa, C., & Otasowie, K. (2024a). Digitisation in the South African construction industry: Recurring challenges. In K. Arai (Ed.), *Proceedings of the Future Technologies Conference (FTC) 2024, Volume 4*. FTC 2024. Lecture Notes in Networks and Systems (Vol. 1157). Springer. https://doi.org/10.1007/978-3-031-73128-0_35

Kubjana, L., Adekunle, P., Aigbavboa, C., Akinradewo, O. (2024b). Analyzing the impact of emotional intelligence on leadership in construction 4.0. In K. Arai (Ed.), *Proceedings of the*

Future Technologies Conference (FTC) 2024, Volume 4. FTC 2024. Lecture Notes in Networks and Systems (Vol. 1157). Springer. https://doi.org/10.1007/978-3-031-73128-0_30

Kubota, F. I., Hsuan, J., & Cauchick-Miguel, P. A. (2017). Theoretical analysis of the relationships between modularity in design and modularity in production. *International Journal of Advanced Manufacturing Technology, 89*, 1943–1958. https://doi.org/10.1007/s00170-016-9238-4

Kutsch, E., & Hall, M. (2005). Intervening conditions on the management of project risk: Dealing with uncertainty in information technology projects. *International Journal of Project Management, 23*(8), 591–599. https://doi.org/10.1016/j.ijproman.2005.06.009

Lakshmi, C. L., & Appa Iyer, S. (2013). Application of queueing theory in health care: A literature review. *Operations Research for Health Care, 2*(1–2), 25–39. https://doi.org/10.1016/j.orhc.2013.03.002

Leigh, J. R. (2004). *Control theory* (Vol. 64). IET.

Li, H., Arditi, D., & Wang, Z. (2013). Transaction-related issues and construction supply chain performance. *Journal of Construction Engineering and Management, 139*(5), 04013005.

Li, G.-Q., Cao, K., Lu, Y., & Jiang, J. (2017). Effective length factor of columns in non-sway modular steel buildings. *Advanced Steel Construction, 13*(4), 412–426.

Liang, L. Y., & Chao, W. C. (2008). The strategies of tabu search technique for facility layout optimization. *Automation in Construction, 17*(6), 657–669. https://doi.org/10.1016/j.autcon.2008.01.001

Liew, J. Y. R., Chua, Y. S., & Dai, Z. (2019). Steel concrete composite systems for modular construction of high-rise buildings. *Structures, 21*(February), 135–149. https://doi.org/10.1016/j.istruc.2019.02.010

Lim, W. M. (2022). The sustainability pyramid: A hierarchical approach to greater sustainability and the United Nations Sustainable Development goals with implications for marketing theory, practice, and public policy. *Australasian Marketing Journal, 30*(2), 142–150. https://doi.org/10.1177/18393349211069152

Linnenluecke, M. K., & Griffiths, A. (2010). Corporate sustainability and organizational culture. *Journal of World Business, 45*(4), 357–366.

Liu, H., Al-Hussein, M., Lu, M., & Yu, H. (2016). BIM-based integrated framework for detailed cost estimation. *Automation in Construction, 71*, 80–93.

Lu, W., & Korman, T. (2010). Implementation of building information modeling using virtual prototyping technologies. *Automation in Construction, 19*(2), 158–172.

Lu, W., Xu, J., & Söderlund, J. (2018). Building information modeling and organizational change. *Construction Management and Economics, 36*(11), 623–637.

MacDuffie, J. P. (2013). Modularity-as-property, modularization-as-process, and 'modularity'-as-frame: Lessons from product architecture initiatives in the global automotive industry. *Global Strategy Journal, 3*(1), 8–40. https://doi.org/10.1111/j.2042-5805.2012.01048.x

Mallieswari, R. (2020). Application of simulation in optimizing online sales check out operation in restaurant using tora. *TRANS Asian Journal of Marketing & Management Research, 9*(7 & 8), 4–17. https://doi.org/10.5958/2279-0667.2020.00032.2

Mao, C., Shen, Q., Shen, L., & Tang, L. (2013). Comparative study of greenhouse gas emissions between off-site prefabrication and conventional construction methods: Two case studies of residential projects. *Energy and Buildings, 66*, 165–176.

Mao, W., Wang, W., Sun, H., Yao, P., Wang, X., & Luo, D. (2021). Urban industrial transformation patterns under natural resource dependence: A rule mining technique. *Energy Policy, 156*, Article 112383.

Martin, M. V., & Ishii, K. (2002). Design for variety: Developing standardized and modularized product platform architectures. *Research in Engineering Design, 13*(4), 213–235. https://doi.org/10.1007/s00163-002-0020-2

McKay, S. (2010). *A control theory based hybrid architecture to anticipate and shape adversarial behavior*. Purdue University.

Meel, A., & Seider, W. D. (2006). Plant-specific dynamic failure assessment using Bayesian theory. *Chemical Engineering Science, 61*(21), 7036–7056. https://doi.org/10.1016/j.ces.2006.07.007

Mehandiratta, R. (2011). Applications of queuing theory in health care. *International Journal of Computing and Business Research, 2*(2), 2229–6166.

Mikkola, J. H. (2007). Management of product architecture modularity for mass customization: Modeling and theoretical considerations. *IEEE Transactions on Engineering Management, 54*(1), 57–69. https://doi.org/10.1109/tem.2006.889067

Molavi, J., & Barral, D. L. (2016). A construction procurement method to achieve sustainability in modular construction. *Procedia Engineering, 145*, 1362–1369. https://doi.org/10.1016/j.proeng.2016.04.201

Momani, A. M., & Jamous, M. (2017). The evolution of technology acceptance theories (April 7, 2017). *International Journal of Contemporary Computer Research (IJCCR), 1*(1), 51–58. Available at SSRN: https://ssrn.com/abstract=2971454

Monahan, J., & Powell, J. C. (2011). An embodied carbon and energy analysis of modern methods of construction in housing: A case study using a lifecycle assessment framework. *Energy and Buildings, 43*(1), 179–188.

Mostafa, S., Kim, S., Tam, V. W. Y., & Rahnamayiezekavat, P. (2016). Exploring barriers to implementing modular construction. *Journal of Cleaner Production, 147*, 190–201.

Nadim, W., & Goulding, J. S. (2011). Offsite production: A model for building down barriers: A European construction industry perspective. *Engineering, Construction and Architectural Management, 18*(1), 82–101. https://doi.org/10.1108/09699981111098702

Naqvi, D., Wey, E., Morgan, J., Miller, M., & Nguyen, T. (2014). Transportation considerations in module design. In *Proceedings of the Structures Congress 2014*. Structural Engineering Institute of ASCE.

Nielsen, K. S. (2017). From prediction to process: A self-regulation account of environmental behavior change. *Journal of Environmental Psychology, 51*, 189–198. https://doi.org/10.1016/j.jenvp.2017.04.002

O'Connor, J. T., O'Brien, W. J., & Choi, J. O. (2014). Critical success factors and enablers for optimum and maximum industrial modularization. *Journal of Construction Engineering and Management, 140*(6), 1–11.

O'Connor, J. T., O'Brien, W. J., & Choi, J. O. (2015). Standardization strategy for modular industrial plants. *Journal of Construction Engineering and Management, 141*(9), 1–10. https://doi.org/10.1061/(asce)co.1943-7862.0001001

Ogata, I. (2002). A proof theoretical account of continuation passing style. In *Computer Science Logic: 16th International Workshop, CSL 2002 11th Annual Conference of the EACSL*, Edinburgh, Scotland, UK, September 22–25, 2002 Proceedings 16 (pp. 490–505). Springer.

Otasowie, K., Aigbavboa, C., Oke, A., & Adekunle, P. (2024). Factors influencing circular economy adoption in the South African construction sector. In J. O. B. Rotimi, W. M. Shahzad, M. Sutrisna, & R. Kahandawa (Eds.), *Advances in Engineering Management, Innovation, and Sustainability*. EPPM 2023. Lecture Notes in Civil Engineering (Vol. 480). Springer. https://doi.org/10.1007/978-3-031-56544-1_43

Oyama, K., Hernádi, I., Iijima, T., & Tsutsui, K.-I. (2010). Reward prediction error coding in dorsal striatal neurons. *The Journal of Neuroscience, 30*(34), 11447–11457. https://doi.org/10.1523/jneurosci.1719-10.2010

Oztemel, E., & Gursev, S. (2020). Literature review of Industry 4.0 and related technologies. *Journal of Intelligent Manufacturing, 31*, 127–182. https://doi.org/10.1007/s10845-018-1433-8

Pajares, F. (2002). Gender and perceived self-efficacy in self-regulated learning. *Theory into Practice, 41*(2), 116–125. https://doi.org/10.1207/s15430421tip4102_8

Pan, W., Gibb, A. G., & Dainty, A. R. (2012). Strategies for integrating the use of off-site production technologies in house building. *Journal of Construction Engineering and Management*, 1331–1340. https://doi.org/10.1061/(ASCE)CO.1943-7862.0000544

Pan, W., & Goodier, C. (2012). House-building business models and off-site construction takeup. *Journal of Architectural Engineering, 18*, 84–93. https://doi.org/10.1061/(ASCE)AE.1943-5568.0000058

Pang, S. D., Liew, J. Y. R., Dai, Z., & Wang, Y. (2016). Prefabricated prefinished volumetric construction joining techniques review. In *Proceedings of the Modular and Offsite Construction Summit*, Edmonton, Canada, pp. 249–256 (2016)

Papazoglou, M. P., & Van Den Heuvel, W. J. (2006). Service-oriented design and development methodology. *International Journal of Web Engineering and Technology, 2*(4), 412–442.

Pasquire, C., Gibb, A., & Blismas, N. (2004). Off-site production: Evaluating the drivers and constraints. In *Proceedings 12th Annual Conference of the International Group for Lean Construction (IGLC12)*. IGLC.

Peltokorpi, A., et al. (2018). Categorizing modularization strategies to achieve various objectives of building investments. In *Construction Management and Economics*.

Peltokorpi, A., Olivieri, H., Granja, A. D., & Seppänen, O. (2017). Categorizing modularization strategies to achieve various objectives of building investments. *Construction Management and Economics, 36*(1), 32–48. https://doi.org/10.1080/01446193.2017.1353119

Pervez, H., Ali, Y., & Pamucar, D. (2021). Risk assessment of modular construction supply chains. *Sustainability, 13*(10), 5311.

Pervez, H., Ali, Y., Pamucar, D., Garai-Fodor, M., & Csiszárik-Kocsir, Á. (2022). Evaluation of critical risk factors in the implementation of modular construction. *PLOS ONE, 17*(8), e0272448. https://doi.org/10.1371/journal.pone.0272448

Phulwani, P. R., Kumar, D., & Goyal, P. (2020). A systematic literature review and bibliometric analysis of recycling behavior. *Journal of Global Marketing, 33*(5), 354–376.

Pour Rahimian, F., Seyedzadeh, S., Oliver, S., Rodriguez, S., & Dawood, N. (2020). On-demand monitoring of construction projects through a game-like hybrid application of BIM and machine learning. *Automation in Construction, 110*, Article 103012. https://doi.org/10.1016/j.autcon.2019.103012

Prothero, A., Dobscha, S., Freund, J., Kilbourne, W. E., Luchs, M. G., Ozanne, L. K., & Thøgersen, J. (2011). Sustainable consumption: Opportunities for consumer research and public policy. *Journal of Public Policy & Marketing, 30*(1), 31–38.

Qi, B., Razkenari, M., Costin, A., Kibert, C., & Fu, M. (2021). A systematic review of emerging technologies in industrialized construction. *Journal of Building Engineering, 39*(February), Article 102265. https://doi.org/10.1016/j.jobe.2021.102265

Quale, J., Eckelman, M. J., Williams, K. W., Sloditskie, G., & Zimmerman, J. B. (2012). Construction matters: Comparing environmental impacts of building modular and conventional homes. *Journal of Industrial Ecology, 16*(2), 243–253.

Ramaji, I. J., & Memari, A. M. (2013). Identification of structural issues in design and construction of multi-story modular buildings. In *Proceedings of the 1st Residential Building Design and Construction Conference* (pp. 294–303).

Ramaji, I. J., Memari, A. M., & Messner, J. I. (2017). Product-oriented information delivery framework for BIM-enabled design. *Automation in Construction, 83*, 264–277.

Razkenari, M., Fenner, A., Shojaei, A., Hakim, H., & Kibert, C. (2020). Perceptions of offsite construction in the United States: An investigation of current practices. *Journal of Building Engineering, 29*(December 2019), 101138. https://doi.org/10.1016/j.jobe.2019.101138

Sarmidi, T., Hook Law, S., & Jafari, Y. (2013). Resource curse: New evidence on the role of institutions. *International Economic Journal, 28*(1), 191–206.

Saurin, T. A., Rooke, J., & Koskela, L. (2013). A complex systems theory perspective of lean production. *International Journal of Production Research, 51*(19), 5824–5838. https://doi.org/10.1080/00207543.2013.796420

Salonitis, K., Pandremenos, J., Paralikas, J., & Chryssolouris, G. (2010). Multifunctional materials: Engineering applications and processing challenges. *International Journal of Advanced Manufacturing Technology, 49*(9–12), 803–826. https://doi.org/10.1007/s00170-009-2428-6

Schoenwitz, M., Naim, M., & Potter, A. (2012). The nature of choice in mass customized house building. *Construction Management and Economics, 30*(3), 203–219.

Schuh, G., Arnoscht, J., & Rudolf, S. (2010). Integrated development of modular product platforms. In *Proceedings of PICMET 2010: Technology Management for Global Economic Growth* (pp. 1–13). IEEE.

Seshaiah, C., & Thiagaraj, H. (2011). A queueing network congestion model in hospitals. *European Journal of Scientific Research, 63*(3), 419–427.

Shahin, A., & Dabestani, R. (2011). A feasibility study of the implementation of total quality management based on soft factors. *Journal of Industrial Engineering and Management, 4*(2), 258–280.

Shahtaheri, Y., Osanloo, M., & Karimi, B. (2015). A multi-criteria decision-making approach to modular system selection. *International Journal of Mining Science and Technology.*

Shahtaheri, Y., Osanloo, M., & Karimi, B. (2017). Modular system design under uncertainty. *Resources Policy, 52*, 66–77.

Smith, R. E. (2014). Off-site construction implementation resource: Off-site and modular construction explained. National Institute of Building Sciences. https://www.wbdg.org/resources/offsiteconstructionexplained.php

Singer, M., & Donoso, P. (2008). Assessing an ambulance service with queueing theory. *Computers and Operations Research, 35*, 2549–2560.

Smith, R., & Rice, T. (2015). *Permanent modular construction: Process, practice, performance.* Modular Building Institute.

Smith, R., & Rice, T. (2017). Chapter 7: Permanent modular construction. In *Offsite architecture: Constructing the future.*

Steg, L., & Vlek, C. (2009). Encouraging pro-environmental behaviour: An integrative review and research agenda. *Journal of Environmental Psychology, 29*(3), 309–317.

Taherdoost, H. (2018). Development of an adoption model to assess user acceptance of e-service technology: E-service technology acceptance model. *Behaviour & Information Technology, 37*(2), 173–197. https://doi.org/10.1080/0144929x.2018.1427793

Taherdoost, H. (2021). Importance of technology acceptance assessment for successful implementation and development of new technologies. *Global Journal of Engineering Sciences.*

Tan, J., Lo, K., Qiu, F., Liu, W., Li, J., & Zhang, P. (2017). Regional economic resilience: Resistance and recoverability of resource-based cities during economic crises in northeast China. *Sustainability, 9*(12), 2136.

Taylor, A., Cocklin, C., Brown, R., & Wilson-Evered, E. (2009). An investigation of champion-driven leadership processes. *Leadership & Organization Development Journal, 30*(4), 345–362.

Ulrich, K. T. (1995). The role of product architecture in the manufacturing firm. *Research Policy, 24*(3), 419–440.

Velamati, S. (2012). Conference report: State of the art modular construction in 2017.

Wang, J., Liu, B., & Li, J. (2008). Transient analysis of an M/G/1 retrial queue subject to disasters and server failures. *European Journal of Operational Research, 189*(3), 1118–1132.

Weissmann, M. A. (2020). *A behavioural control investigation towards greater sustainability.* Swinburne University of Technology.

Wong, J. K. W., & Lau, E. (2015). From the traditional procurement to the BIM-integrated procurement. *Journal of Facilities Management, 13*(4), 364–382.

Wuni, I. Y., & Shen, G. Q. (2020). Barriers to the adoption of modular integrated construction: Systematic review and meta-analysis, integrated conceptual framework, and strategies. *Journal of Cleaner Production, 249*, Article 119347. https://doi.org/10.1016/j.jclepro.2019.119347

Xie, J. (2015). Preparing for our prefab future. *Curbed.*

Yang, G., Miao, G., Zhang, X., et al. (2023). Research on value creation of engineering project from the perspective of knowledge creation fusion—Based on stakeholder theory. *Journal of the Knowledge Economy, 14*, 2723–2741. https://doi.org/10.1007/s13132-022-00963-6

Yarlagadda, P. K. D. V., Cheng, K., & Annamalai, A. R. (2017). Platform-based manufacturing systems. *Procedia Manufacturing, 10*, 94–105.

Zhai, X., & Reed, R. (2014). Off-site construction and sustainability. *Journal of Architectural Engineering, 20*(2), 04013012.

Zhang, P., Zhang, X., Luiu, J., & Chen, Z. (2016). Structural design and analysis of multi-storey steel structure module and steel frame composite structures. *Build Struct, 46*(10), 95–100.
Zhong, Z., Zhang, Y., Zhang, J., & Su, M. (2024). How information and communication technologies contribute to rural tourism resilience: Evidence from China. *Electronic Commerce Research*. https://doi.org/10.1007/s10660-024-09854-1

Chapter 4
Gaps in Modularity Methodology Theoretical Frameworks

Abstract This chapter looks at the deficiencies of the theoretical frameworks for modularity methodology in construction that are currently in use, emphasising important areas that are either poorly understood or not sufficiently covered. Even though the modularity methodology has made great strides in design, manufacturing, and assembly processes, some aspects, most notably reassembly, maintenance, and cultural perception, require more thorough theoretical integration to increase its efficacy and versatility. Reassembly is the first topic covered in the discussion, bringing to light the lack of theoretical focus on the methodical disassembly and repurposing of modular components. This feature is essential for increasing the lifespan of modular buildings, decreasing waste, and advancing circular economy principles. The chapter then moves on to maintenance, highlighting the necessity of frameworks that integrate modular design and planning with predictive, preventive, and corrective maintenance techniques. Modular buildings may lose their long-term performance, safety, and cost-effectiveness if this factor is not taken into account. Another important disparity is cultural perception, since user acceptance, industry customs, and societal attitudes all have a big impact on the uptake and success of modular construction. Social and cultural barriers that can impede implementation, such as reluctance to use non-traditional building techniques or false beliefs about their quality and durability, are often overlooked by many theoretical models currently in use. The chapter emphasises the value of creating more thorough theoretical frameworks that incorporate technical, operational, and sociocultural aspects by recognising and evaluating these gaps. Modularity methodology will be strengthened, its sustainability potential increased, and its acceptance in a variety of contexts improved by addressing these neglected areas. Future investigations and creative solutions targeted at addressing these significant theoretical flaws will be built upon this analysis.

P. Adekunle et al., *Modularity Methodology*,
https://doi.org/10.1007/978-3-032-12918-5_4

4.1 Introduction

Modularity methodology is widely used in the construction industry to measure the degree of modularity in building projects (Wong & Loo, 2022). However, several gaps in the conceptual framework of modularity methodology can limit their effectiveness in practice. One significant conceptual factor is the lack of standardisation in the terminology and definitions used to describe modularity in construction. Different methodologies use various terms and definitions, making it difficult to compare and analyse data across different projects (Musa et al., 2014). This lack of standardisation can limit the effectiveness of modularity methodology, leading to inconsistent results and making it difficult to draw meaningful conclusions.

Another factor in the conceptual framework of modularity methodology is its limited scope (Fan et al., 2018). Most methodologies focus on using standardised building components in construction but do not account for other factors that can impact the degree of modularity in a building project. For example, they may not consider the integration of building systems, the use of digital technologies, or the management of the construction process. This can limit their usefulness in practice as they may not fully capture the complexity of modularity in construction. Insufficient data is another significant gap in the conceptual framework of modularity methodology (Tommelein, 2015). There is a lack of empirical data on the effectiveness of modularity methodology in construction, making it difficult to validate its results and improve its accuracy. Evaluating the effectiveness of different methodologies is difficult without sufficient data and making evidence-based decisions about which methodologies to use in different contexts is challenging.

Maintenance is another factor in the conceptual framework of modularity methodology (Dong et al., 2015). The degree of modularity in a building project is highly context-specific and can vary depending on factors such as project requirements, construction methods, and manufacturing capabilities. Modularity methodology may not be able to account for all of these maintenance factors, which can limit its effectiveness in practice. Finally, there is a lack of integration between modularity methodology and other project management tools and techniques. Modularity methodologies are often used in isolation without considering their integration with other project management tools and techniques (Lu et al., 2018). This can limit their practical usefulness, as they may not fully integrate into the overall project management framework. These gaps in the conceptual framework of modularity methodology can limit their effectiveness in practice.

This section of the study identifies the gaps in the conceptual framework of modularity methodology. There are likely some gaps in the Western conceptual framework that have failed to capture the factors affecting modularity methodology in developing countries, because the conceptual framework provides the perspectives from which problems are highlighted. Addressing the gaps will require a more comprehensive and standardised approach to measuring modularity in construction and more empirical research on the effectiveness of different methodologies in different contexts. To help construction professionals optimise the construction process and enhance project outcomes, a more effective modularity methodology might be developed by filling in these gaps.

4.2 Reassembly

Reassembly involves a collection of related parts or types to enable the reuse of standardised design or specification pieces, enhancing design and delivery productivity and providing a place to store specifications and lessons learned understandably and practically (Gallego-Schmid et al., 2020). The related parts might have comparative statistics on costs and carbon effects. Assembly components themselves may include characteristics and classifications. These attributes might include crucial information such as cost, CO_2, programme, maintenance, and other crucial details attached to the object for usage once it is put into a model. Therefore, reassembly could involve putting building components back together. For instance, a building's frame might have been pre-assembled in a workshop and then disassembled before being shipped to the site, only to be permanently reassembled there (Coelho & Brito, 2012). Reassembly in construction is disassembling current buildings or structures and utilising their parts to build new ones. As a greener option to conventional building and demolition techniques, the reassembly strategy is gaining popularity in the construction sector.

Reassembling entails meticulously tearing down an existing structure or building, recovering the salvageable parts, and using them in brand-new construction projects (Glias, 2013). By minimising waste and the demand for new building materials, this strategy can mitigate the adverse environmental effects of construction projects. For historic or architecturally significant constructions where it is crucial to preserve the original components, reassembly is especially advantageous. Construction experts can maintain the building's historic charm while producing a practical and contemporary environment by recovering and reusing the original components (Barrett & Scott, 2012). Reassembly has advantages for the environment and the past; however, it can also reduce the cost of construction projects. Construction workers can cut the cost of new building materials and the time and labour necessary for new construction by reusing old components. This may result in shorter project timelines and lower building costs.

The reassembly procedure, however, is not without its difficulties. For instance, disassembly can be labour and time-intensive, while transferring salvaged materials to new construction sites may provide logistical difficulties (Heyes et al., 2018). Utilising salvaged parts may also raise structural or safety issues that must be appropriately considered and resolved. Despite these difficulties, reassembly is becoming increasingly popular as a sustainable and economical replacement for conventional building techniques. Construction experts can develop contemporary, functional buildings while retaining the historic character of the original structure and reducing the environmental impact of construction projects by carefully demolishing and repurposing existing structures.

4.2.1 Constructivist Theory

Constructivism is a learning theory suggesting that individuals construct knowledge and meaning based on their experiences and interactions with the world (Juvova et al., 2015). This theory proposes that learning is an active, social process that involves the learner's experiences and beliefs, and that knowledge is not simply transmitted but constructed by the learner. In the context of construction, constructivism can be applied in various ways. One way is through project-based learning, which is an approach that allows learners to engage actively in the construction process and learn by doing (Chudy & Neumeister, 2011). Through hands-on experience, learners can better understand construction concepts and principles and develop problem-solving skills and critical thinking abilities. Another way constructivism can be applied in construction is through collaborative learning. This involves learners' working in groups to solve problems, share ideas, and learn from each other (Burner, 2003). In construction, collaborative learning can be applied through group projects, where learners work together to design and build a structure, and through peer review, where learners provide feedback and constructive criticism to one another.

Constructivism can also be applied in construction through technology (Hejny et al., 2004). Virtual and augmented reality technologies can provide learners with immersive learning experiences where they can interact with a virtual construction site and learn by doing. This can help learners better understand construction concepts and principles and prepare them for real-world construction scenarios. Constructivism offers a valuable approach to learning in the construction industry (Skoda & Doulik, 2009). Construction professionals may create meaningful, interactive, and engaging learning experiences by acknowledging that learners develop knowledge and meaning via their experiences and relationships. This can support students' acquisition of the abilities and information required for success in the construction sector and assist the sector as a whole to continue to expand and develop.

4.2.2 Schema Theory

A cognitive theory known as schema theory contends that people process and interpret incoming information using pre-existing knowledge frameworks known as schemas (An, 2013). Schemas are conceptual frameworks that people create owing to their experiences, education, and worldviews. These frameworks make it simple to assess and react immediately to new information by comparing it to their pre-existing schemas. Schema theory applies to building in a variety of ways. Utilising diagrams and visual aids is one method. Learners can receive and comprehend new material more quickly by being given visual representations of construction principles and concepts to compare them to their preexisting schemas and see how they fit (Van Kesteren et al., 2014). This can enhance learners' capacity to apply their knowledge in practical situations and help them better understand construction concepts and principles.

Case studies are another method of using schema theory in building (Plant & Stanton, 2012). Educators can assist students in developing mental models that they can use to take in and understand new information by providing real-world examples of building projects. By applying their schemas to new circumstances, students can refine their critical thinking and problem-solving abilities by selecting the optimal course of action. In addition, simulators can be used to implement schema theory in the building industry (Østergård et al., 2016). Construction stakeholders can create schemas based on their experiences and expertise by being presented with simulated building scenarios. They can then utilise these schemas to process and interpret new information. Simulations can also give participants a safe, controlled environment to experiment with various construction methods and procedures and better understand how these methods function (Harris et al., 2020). Generally, schema theory offers a valuable teaching method in the building sector. Construction industry experts may create learning experiences that are engaging, dynamic, and meaningful by acknowledging that people process and interpret new information using pre-existing knowledge structures. This can assist construction professionals in gaining the abilities and information necessary to prosper in the construction sector and contribute to the continued expansion and development of the sector.

4.2.3 Social Constructivism

A learning theory called social constructivism strongly emphasises how social and cultural contexts affect knowledge creation (Amineh & Asl, 2015). According to this concept, knowledge is created through social contact and speech and is a collaborative, dynamic process that requires social bargaining. Social constructivism emphasises the role of language and communication in knowledge production and holds that social and cultural settings shape knowledge. Social constructivism has several applications in the field of construction. The use of collaborative learning is one approach. Professionals collaborating in groups can exchange concepts, clarify meaning, and build knowledge through social contact and dialogue (Adams, 2006). As a result, professionals may gain a deeper comprehension of construction-related ideas and principles and improve problem-solving and critical thinking skills. Apprenticeships and mentorships are other methods of social constructivism that can be used in the building industry. Working alongside seasoned construction professionals allows students to benefit from their knowledge and experiences while gaining a greater understanding of the sector (Bereiter, 2001). This can assist construction stakeholders in gaining the abilities and information necessary to thrive in the field and gain a sense of self and community within the construction business.

Using community-based learning, social constructivism can also be used in buildings (Easen & Bolden, 2005). By interacting with local communities and stakeholders, learners can gain a deeper grasp of the social and cultural influences on construction practices and processes. As a result, students may have a more comprehensive understanding of the construction sector and be more equipped to collaborate with many stakeholders in actual building projects (MacGilchrist, 2003). Social constructivism offers a valuable teaching method in the building sector. Construction professionals may create engaging, interactive, and meaningful learning experiences by being aware of the social and cultural influences on knowledge construction. This can assist students in gaining the abilities and information necessary to prosper in the construction sector and contribute to the continued expansion and development of the sector.

4.3 Maintenance

Maintenance ensures that buildings and other assets retain an excellent appearance and operate optimally (Xian Wei & Syahrul Mat Radzuan, 2021). Inadequate maintenance can result in decay, degradation, and reduced performance, as well as affect the health and threaten the safety of users, occupants, and others in the vicinity. Depending on their design, quality of materials and artistry, function and location, buildings deteriorate at different rates and require different levels of attention. No building will ever be maintenance-free; however, the design and artistry quality can minimise the required level (Ofori et al., 2015). The growing significance of building maintenance has also generated an increasing interest in developing maintenance management procedures to improve the performance of buildings. As an illustration, it was concluded that the fundamental problem is ineffective maintenance management. Yacob et al. (2016) also observed a need to review, evaluate, and reassess maintenance policies with serious consideration for safety and security criteria. Kayan (2006) outlined that there is a need for proper maintenance management in order to minimise building maintenance.

Ling et al. (2014) also outlined that maintenance management is inconclusive as it does not consider risk factors while formulating maintenance policy. Based on the outcome of the research conducted by Syamilah (2006), more than 90% of the respondents concurred that there is a need for further improvement (which should be broader) in the practice of the maintenance management procedure of buildings. Syamilah (2006) further asserted that the dominant maintenance practices are reactive and condition-based. Ahmad and Kamaruddin (2012) asserted that a fundamental problem with building maintenance problems is the lack of sufficient attention to the maintenance factor. Ignoring or not paying sufficient attention to the maintenance factor of a building will result in a building that will be expensive to own, since more than 50% of the costs of ownership are attributable to maintenance. Research indicates that due to weak maintenance policies, many buildings fail to meet users' expectations (Hauashdh et al., 2022). Studies also show that school buildings suffer from poor management practices. Khalid et al. (2019) highlighted that building maintenance procedures are generally ineffective and largely reactive. Also, the study discussed challenges faced by property managers in high-rise condominiums, which revealed frequent tenant complaints, including delays in attending to defects, poor workmanship, and inadequate facilities and services even at the time of handover from developers. In response, Dzulkifli et al. (2021) recommended a review of existing defect management processes. Consistently emphasising that maintenance practices in Malaysia remain predominantly reactive and condition-based, noting that maintenance factors are not properly managed despite the substantial investments made in multi-storey housing projects.

Many buildings require maintenance in Malaysia; the list seems endless despite the vast amounts each successive government has invested in the sector (Shaikh et al., 2017). Therefore, there is a deficiency in how a building's maintenance procedures are managed. Various attempts have been made to improve the performance of buildings through maintenance. While such procedures offer the potential to improve the performance of maintenance management systems, the systems have, however, been reactive, hypothetical, and conditionally based. These substantial weaknesses in the proposed procedures have created fundamental problems with the existing and proposed building maintenance management procedures, leading to their inability to improve the existing systems. Yun and Choi (2025) stated that maintenance cannot be circumvented. However, it is possible that expenditure on building maintenance can be optimised through a proactive maintenance management system based on the concept of value. Users measure the performance of their building in terms of various criteria consistent with their value systems. Maintenance management procedures must be based on the user's value systems. A significant impetus of value-based maintenance management is the progressive realisation that maintenance must be viewed from engineering, scientific, technological, political, and commercial perspectives.

4.3.1 Maintenance Theory of Reliability

The maintenance of systems in construction is of increasing concern to managers and designers in all branches of engineering, whether they are producing high-quality products or designing highly reliable systems (Waghmare et al., 2014). In recent decades, reliability theory has produced many innovations in maintenance policy. Maintenance theory of reliability surveys useful and practical maintenance models covering replacement, preventive maintenance and inspection (Nannikar et al., 2012). The theory deals with the effect of mean time to repair upon overall system failure rates. However, these calculations are not necessary for critical systems as a crucial performance criterion deal with operational failures, distinct from hazardous failures in that they are the outcome of the system's reaction to prevent unsafe failures. Reliability theory has grown from the valuable experiences of many defects of military systems in World War II and modern technology's development (Kanwarpreet & Ahuja, 2012). The theory has been applied to industrial, mechanical, and electronic engineering, as well as to computer information, computer engineering, and natural systems phenomena, to improve reliability. Many serious accidents have happened in the world where systems have been large scale and complex; these not only cause heavy damage and a social sense of instability but also bring an unrecoverable negative influence on the living environment. These are said to have occurred from various sources of equipment deterioration and maintenance reduction due to a policy of industrial rationalisation and personnel cuts.

Construction development observers may worry that big earthquakes might soon happen, destroying large old plants, such as chemical and power plants, and severely damaging large areas (Ahmad et al., 2012). Most industries refrain from investing in new plants but try to run current plants safely and efficiently as long as possible. Furthermore, advanced nations with massive infrastructural developments are in the maintenance period of their investments (Moradi et al., 2011). Since maintenance is more vital than redundancy and production, maintenance policies shaped by the maintenance theory of reliability are being applied to natural systems to carry out timely maintenance.

4.3.2 Building Maintenance Management Theory of Reliability

Building maintenance management theory coordinates activities designed to maintain, repair, and improve buildings and their related systems and cost-effectively provide a safe, habitable, comfortable, and functional environment (Shohet & Nobili, 2017). It encompasses all tasks that make a space 'liveable' and ensures that significant building systems work efficiently, such as electrical, plumbing andfire prevention. Maintenance management theory also covers a building's structure, including flooring, walls, ceilings, roofs, and fixtures. In addition, building maintenance may extend to building exteriors and include painting, cleaning, landscaping, and groundskeeping. Energy management is essential to building maintenance, especially in ageing structures (Jones & Sharp, 2007). Proper energy management leads to greater enjoyment of building spaces, better health and living conditions for occupants, and lower energy bills. Therefore, building maintenance teams must inspect and track the condition of roofs, insulation, doors and windows, and other areas that might contribute to the loss of heat or cold.

Proper building maintenance management is vital for the safety and comfort of occupants, whether residents, employees, or commercial tenants (Cingolini et al., 2008). Though occupants are responsible for maintenance tasks, such as changing light bulbs and maintaining a sanitary environment, they become disgruntled when common areas are not cleaned or when safety issues arise. Furthermore, ignored maintenance issues accumulate over time, creating dangerous environments, sometimes with tragic results. In addition, prospective occupants need assurance from their lessors or owners that their maintenance concerns will be addressed promptly (Lavy et al., 2014). It is much easier for building owners to manage a maintenance budget with predictable costs than to try to find the money whenever issues occur. For example, replacing a significant building system is costlier than proactively maintaining building assets through regular preventive maintenance (PM). Well-maintained buildings are typically more energy efficient, leading to further cost savings. Adequate maintenance is essential to remain in compliance with local, state, and federal regulations, such as those imposed by the Occupational Safety and Health Administration (OSHA), the National Fire Protection Association (NFPA), and the International Code Council (ICC). Building code violations can result in heavy fines and penalties for building owners. Remaining compliant with building codes also helps minimise safety risks and reduces liabilities to occupants or guests who use the building.

4.3.3 Asset Management Theory

A methodical approach to managing physical assets throughout their existence, from acquisition to disposal, is known as asset management theory (Van Riel et al., 2016). The management of the lifecycle of construction assets, including buildings, bridges, and other infrastructure, can be accomplished using this approach, which is widely employed in various industries, including construction. Asset management theory in construction encompasses a variety of tasks, including planning for maintenance, performance evaluation, risk management, and decision-making. Asset management in the construction industry aims to maximise asset value throughout its lifecycle while minimising costs and hazards (Dirksen et al., 2013). A comprehensive strategy that considers the asset's whole lifecycle, from design to decommissioning, is necessary for effective asset management in the construction industry. This includes the asset's operational, financial, and legal components and physical features.

Maintenance planning is a crucial component of asset management theory in construction. In order to guarantee the longevity, effectiveness, and safety of building assets, proper maintenance is essential (Van Riel et al., 2014). The asset management theory offers a framework for formulating and executing maintenance plans founded on facts and analysis rather than conjecture. Risk management is another critical component of asset management theory in construction. Various dangers, such as natural disasters, equipment malfunctions, and vandalism, can affect construction assets (Kleidorfer et al., 2013). Effective asset management entails recognising and assessing these risks, creating mitigation measures, and creating contingency plans to guarantee that assets are safeguarded and operable in the event of an unforeseen incident. Asset management theory is a valuable method for managing construction assets throughout their existence. Construction professionals may maximise asset value and reduce costs and risk by using a systematic and comprehensive approach to asset management.

4.4 Cultural Perception

Cultural perception describes how individuals from various cultural backgrounds view and comprehends their surroundings (Teravainen et al., 2018). Cultural perception can significantly affect the construction process in the context of the building industry, affecting everything from design and planning to communication and collaboration among project stakeholders. The process of design and planning is one of the most critical ways that cultural perception affects building. The preferences of other cultures for building materials, designs, and features may differ, impacting the design decisions made by architects and engineers (Koh & Low, 2010). For instance, certain cultures may place more importance on aesthetics or durability, while others may consider energy efficiency and sustainability more. Buildings and structures that satisfy the needs and expectations of various groups must consider these cultural preferences.

The way project stakeholders interact and engage with one another is also influenced by cultural perspectives (Lückmann & Färber, 2015). For instance, communication difficulties among team members from diverse cultural origins can arise due to language issues. This is similar to how differing cultural norms around hierarchy, decision-making, and communication can influence how project teams interact with one another. When it comes to addressing problems or making decisions on construction projects, for instance, it may be improper in some cultures to question or criticise authority. During the actual construction process, cultural perspectives also impact the building. The safety, effectiveness, and quality of construction projects may be impacted by diverse cultural construction styles or approaches (Trinh & Feng, 2020). For instance, whereas some cultures stress using machines and modern technology, others may place a more significant premium on manual labour and conventional building techniques.

The use and perception of buildings and structures by their occupants may be influenced by cultural perception (Mahdavi, 2020). Building use and maintenance practices may change depending on the culture, impacting the structure's long-term sustainability and longevity. To ensure that buildings and structures are designed and built in a way that fits the needs and expectations of their users, it is crucial to understand these cultural perspectives. In summary, cultural perception significantly impacts the construction sector, affecting everything from design and planning to stakeholder communication and collaboration (Adekunle et al., 2021). Construction professionals may design buildings and structures that suit the needs and expectations of various populations while guaranteeing safety, efficiency, and quality throughout the construction process by acknowledging and respecting these cultural differences.

4.4.1 *Individualism-Collectivism*

According to the individualism-collectivism cultural concept, people value their interests over the interests of the collective to varying degrees (Germani et al., 2020). Collectivistic cultures value social cohesion, loyalty, and interdependence more than individualistic cultures do individual success, autonomy, and independence. The individualism-collectivism paradigm has several applications in the field of building. Understanding cultural variations and how they affect building methods and procedures is one approach to this (Burson et al., 2012). For instance, construction professionals may place their accomplishments and career growth over the requirements of the team or company in individualistic cultures, which can result in a competitive and dispersed work environment. Construction workers may value group harmony and collaboration in collectivist societies, which can result in a more cooperative and collaborative work environment.

The concepts of individualism and collectivism may also be used in the administration of teams and organisations in the construction industry (Schnabel et al., 2013). Construction managers may create team structures and procedures that encourage successful communication and cooperation by having a thorough awareness of each team member's cultural values and objectives. Construction managers may need to set clear guidelines and expectations for individual roles and responsibilities in individualistic cultures, for instance, while prioritising team-building exercises and group decision-making procedures in collectivistic cultures (Germani et al., 2020). Additionally, the philosophy of individualism-collectivism may be used to build and foster intercultural communication abilities. Construction professionals may collaborate on international construction projects and interact successfully with stakeholders from different cultural backgrounds by identifying and comprehending cultural differences. As a result, the construction sector may become more inclusive and varied and operate more efficiently (Copp et al., 2015). The individualism-collectivism paradigm, in general, provides an insightful prism through which to view the cultural influences that affect construction techniques and procedures. Construction professionals may create more successful teams, organisations, and communication strategies by considering cultural variations, which will also help create a more inclusive and varied construction sector.

4.4.2 *Social Identity Theory*

A psychology theory called social identity theory explains how people define themselves through social groupings (Ellemers & Haslam, 2012). This idea contends that people feel self-worth and a sense of identity from belonging to social groupings, such as their race, profession, or sports team. According to social identity theory, intergroup prejudice and discrimination might result from people's motivation to build strong social identities. Social identity theory has a variety of applications in the field of building (Barreto & Ellemers, 2000). One method is the growth of a feeling of identity and belonging inside the building sector. Individuals identifying as construction professionals might feel proud of and satisfied with their jobs and build a sense of purpose and mission with other construction professionals. Building projects may result in higher levels of motivation, cooperation, and productivity.

Managing teams and organisations is another way to incorporate social identity theory into the construction industry (Derks et al., 2007). Construction managers may create teams and procedures that foster shared identity and belonging by acknowledging the significance of social identity and group membership. Construction managers may encourage a feeling of group cohesiveness and commitment among team members, for instance, by highlighting the team's value and acknowledging the team's accomplishments. Social identity theory can be applied to construct inclusive and diverse communities. In order to promote a more varied and inclusive industry, construction workers should recognise the importance of social identity and the negative impacts of intergroup prejudice (Haslam et al., 2010). This may be accomplished by cultivating a climate of mutual respect and cooperation among all construction industry experts and aggressively seeking out and respecting varied ideas and experiences. The relevance of social identity and group membership in the construction business may be better understood using the framework provided by social identity theory (Leach et al., 2008). Construction professionals may contribute to a more collaborative, effective, and inclusive sector by acknowledging and fostering a feeling of shared identity and belonging among their peers and by supporting diversity and inclusion.

4.4.3 Ecocultural and Sociocultural

An area of cultural theory called ecocultural theory explains how the built and natural environments impact human behaviour, social structures, and cultural practices (Nsamenang, 2015). This theory suggests that the interaction between people and their physical surroundings shapes their lifestyle, values, and community dynamics. Ecocultural theory is a branch of cultural theory that describes how the natural and built environment influences human behaviour (Grishaeva et al., 2021). According to ecocultural theory, people develop cultural practices and behaviours adapted to their physical environment and shaped by their interactions. Ecocultural theory suggests that human behaviour is not only influenced by social and cultural factors but also by physical and environmental factors. In the construction context, ecocultural theory can be applied in various ways (Abdulhakova, 2011). One way is to design and construct sustainable and environmentally friendly buildings. Building designers may create structures encouraging sustainable and healthful living by knowing how the built environment influences human behaviour. According to ecocultural theory, people form cultural norms and practices specific to their physical environment and are shaped by their interactions with the physical environment (Yu et al., 2020). According to ecocultural theory, environmental and physical influences can also affect human, social and cultural behaviour. Ecocultural theory has a variety of applications in the field of construction. Buildings that are ecologically friendly and sustainable may be created in several ways. Building designers may create structures that encourage healthy and sustainable living by understanding how the built environment affects human behaviour (Glenberg et al., 2013). Construction experts may, for instance, lower buildings' carbon footprints and encourage sustainable living habits by utilising green building methods and materials.

On the other hand, sociocultural theory explains how social and cultural elements affect human behaviour and education. According to sociocultural theory, people engage with others and their social and cultural environment to learn and improve their cognitive and social abilities (Hutto & Myin, 2013). The relevance of social contact and cultural settings in influencing human behaviour is emphasised by sociocultural theory. There are numerous methods of applying sociocultural theory to the field of building. Buildings representing local culture and values can be designed and built as one method (Wiley, 2019). Buildings may represent the cultural character and legacy of the community by using materials and architectural aspects that are local to the area. This can help community members feel pride and belonging and help keep cultural traditions alive.

The administration of groups and teams may also be used to implement sociocultural theory in the construction industry (Pokrovskaia et al., 2019). Construction managers may create teams and procedures to encourage successful communication and cooperation by understanding how social and cultural variables affect team dynamics. Construction managers may foster a climate of inclusion and teamwork among team members, for instance, by recognising and respecting the various opinions and experiences of team members. Ecocultural and sociocultural theories provide frameworks for comprehending the influence of environmental, social, and cultural aspects on building methods and procedures. Construction professionals may support a more sustainable, diverse, and inclusive sector by acknowledging the impact of these elements and implementing them into construction design, management, and cooperation.

4.5 Summary

The gaps in the modularity methodology research have been identified and explored in this chapter, emphasising the conceptual framework and three critical gaps: the maintenance factor, the reassembly factor, and the cultural perception aspect. These gaps have been analysed via the prism of pertinent theories to offer a thorough grasp of the difficulties and possible solutions in using modularity techniques for the construction sector. In order to address the complexity inherent in modular systems, a more comprehensive and robust approach is required, as shown by the conceptual framework for modularity methodology. The research on modularity methodology has advanced; nevertheless, significant gaps still prevent its full benefits from being realised. Maintenance is essential for modular structures to remain reliable and sustainable over time. Studies have shown that present approaches frequently undervalue the significance of proactive and systematic maintenance procedures. These findings are supported by the maintenance theory of reliability, building maintenance management theory, and asset management theory.

The longevity and effectiveness of modular systems are guaranteed by effective maintenance; nevertheless, to close the gap between theoretical ideas and practical application, this field needs more concentrated study and practical application. The reassembly factor is essential for modular systems to be flexible and adaptable. According to social constructivism, constructivist theory, and schema theory, the cognitive processes involved in reassembling modular components are complicated and impacted by individual and social influences. The cognitive aspects are not sufficiently addressed by current approaches, which results in reassembly errors and inefficiencies. More studies are required to create instruments and methods to improve the reassembly process by utilising these theoretical insights.

Lastly, in many cultural situations, the cultural perception factor dramatically influences the adoption and efficacy of modular systems. It has become clear from studying individualism-collectivism, social identity theory, and eco-cultural and sociocultural theories that cultural views greatly influence how well modular systems are implemented. These cultural factors are frequently overlooked in current research, leading to methods that might not be widely useful or efficient. A greater comprehension of cultural dynamics and the creation of culturally sensitive approaches is needed to close this gap.

References

Abdulhakova, L. R. (2011). Cultural identity and social development. *Psychology in Russia: State of the Art, 4*, 234–246.

Adams, P. (2006). Exploring social constructivism: Theories and practicalities. *Education 3-13, 34*(3), 243–257. https://doi.org/10.1080/03004270600898893

Adekunle, P., Aigbavboa, C., Thwala, W., & Akinradewo, O. (2021, November). Influence of organisational values and beliefs on sustainable construction performance in the South African construction industry. In *CM50 Conference* (p. 74).

Ahmad, M. F., Zakuan, N., Jusoh, A., & Takala, J. (2012). Relationship of TQM and business performance with mediators of SPC, lean production and TPM. *Procedia—Social and Behavioral Sciences, 65*, 186–191.

Ahmad, R., & Kamaruddin, S. (2012). An overview of time-based and condition-based maintenance in industrial application. *Computers & Industrial Engineering, 63*(1), 135–149. https://doi.org/10.1016/j.cie.2012.02.002

Amineh, R., & Asl, H. (2015). Journal of social sciences, literature and languages review of constructivism and social constructivism. *Journal of Social Sciences, Literature and Languages, 1*(1), 9–16.

An, S. (2013). Schema theory in reading. *Theory & Practice in Language Studies (TPLS), 3*(1).

Barrett, J., & Scott, K., (2012). Link between climate change mitigation and resource efficiency: A UK case study. *Global Environmental Change, 22*, 299–307.

Barreto, M., & Ellemers, N. (2000). You can't always do what you want: Social identity and self-presentational determinants of the choice to work for a low-status group. *Personality and Social Psychology Bulletin, 26*(8), 891–906. https://doi.org/10.1177/01461672002610001
Bereiter, C. (2001). Situated cognition and how to overcome it. In J. Collins & D. Cook (Eds.), *Understanding learning: Influences and outcomes* (pp. 71–83). Paul Chapman/Open University Press.
Burner, T. (2003). *Identity and learning in intercultural contexts*. Scandinavian University Press.
Burson, A., Crocker, J., & Mischkowski, D. (2012). Two types of value-affirmation: Implications for self-control following social exclusion. *Social Psychological and Personality Science, 3*, 510–516.
Chudy, S., & Neumeister, P. (2011). Specification of assumptions to outline the category "value" in the curriculum. *Acta Humanita, 8*(1), 6–31.
Cingolini, F., Fedele, L., & Villa, A. N. (2008). Managing facilities under the multi-service result-oriented approach: Some insights coming from the field in Italy. *Production Planning and Control, 19*(4), 321–326. https://doi.org/10.1080/09537280802034075
Coelho, A., & de Brito, J. (2012). Influence of construction and demolition waste management on the environmental impact of buildings. *Waste Management, 32*(3), 532–541. https://doi.org/10.1016/j.wasman.2011.11.011
Copp, J. E., Giordano, P. C., Longmore, M. A., & Manning, W. D. (2015). Identity and behavior across the life course. *Social Psychology Quarterly, 78*(2), 109–131. https://doi.org/10.1177/0190272515585083
Derks, B., Van Laar, C., & Ellemers, N. (2007). Social creativity strikes back: Improving motivation and performance of low-status groups. *Journal of Personality and Social Psychology, 93*(3), 389–403. https://doi.org/10.1037/0022-3514.93.3.389
Dirksen, J., Clemens, F. H. L. R., Korving, H., Cherqui, F., Le Gauffre, P., Ertl, T., Plihal, H., Müller, K., & Snaterse, C. T. M. (2013). The consistency of visual sewer inspection data. *Structure and Infrastructure Engineering, 9*, 214–228.
Dong, Y. H., Jaillon, L., Chu, P., & Poon, C. S. (2015). Comparing carbon emissions of precast and cast-in-situ construction methods—A case study of high-rise private building. *Construction and Building Materials, 99*, 39–53.
Dzulkifli, N., Sarbini, N. N., Ibrahim, I. S., Abidin, N. I., Yahaya, F. M., & Nik Azizan, N. Z. (2021). Review on maintenance issues toward building maintenance management best practices. *Journal of Building Engineering, 44*, Article 102985. https://doi.org/10.1016/j.jobe.2021.102985
Easen, P., & Bolden, D. (2005) Location, location, location: What do league tables really tell us about primary schools? *Education 3-13, 33*(3), 49–56.
Ellemers, N., & Haslam, S. A. (2012). Social identity theory. In P. A. M. Van Lange, A. W. Kruglanski, & E. T. Higgins (Eds.), *Handbook of theories of social psychology* (Vol. 2, pp. 379–398). Sage.
Fan, K., Chan, E. H. W., & Qian, Q. K. (2018). Transaction costs (TCs) in green building (GB) incentive schemes: Gross Floor Area (GFA) concession scheme in Hong Kong. *Energy Policy, 119*, 563–573.
Gallego-Schmid, A., Chen, H. M., Sharmina, M., & Mendoza, J. M. F. (2020). Links between circular economy and climate change mitigation in the built environment. *Journal of Cleaner Production, 260*, Article 121115. https://doi.org/10.1016/j.jclepro.2020.121115
Germani, A., Buratta, L., Delvecchio, E., & Mazzeschi, C. (2020). Emerging adults and covid-19: The role of individualism-collectivism on perceived risks and psychological maladjustment. *International Journal of Environmental Research and Public Health, 17*(10). https://doi.org/10.3390/ijerph17103497
Glenberg, A. M., Witt, J. K., & Metcalfe, J. (2013). From the revolution to embodiment: 25 years of cognitive psychology. *Perspectives on Psychological Science, 8*(5), 573–585. https://doi.org/10.1177/1745691613498098
Glias, A. (2013). *The 'Donor Skelet': Designing with re-used structural concrete elements*. M.Sc. thesis. Delft University of Technology, Delft, the Netherlands.

Grishaeva, E., Fedotova, O., & Kolesnikova, I. (2021). Identity development in sociocultural and educational contexts. *European Journal of Contemporary Education, 10*(2), 455–466. https://doi.org/10.13187/ejced.2021.2.455

Harris, D. J., Bird, J. M., Smart, P. A., Wilson, M. R., & Vine, S. J. (2020). A framework for the testing and validation of simulated environments in experimentation and training. *Frontiers in Psychology, 11*, Article 524804. https://doi.org/10.3389/fpsyg.2020.00605

Haslam, S. A., Reicher, S. D., & Platow, M. J. (2010). *The new psychology of leadership: Identity, influence and power*. Psychology Press.

Hauashdh, A., Jailani, J., Rahman, I. A., & AL-fadhali, N. (2022). Strategic approaches towards achieving sustainable and effective building maintenance practices in maintenance-managed buildings: A combination of expert interviews and a literature review. *Journal of Building Engineering, 45*, 103490. https://doi.org/10.1016/j.jobe.2021.103490

Hejny, M., Novotna, J., & Stehlikova, N. (2004). *Twenty-five chapters of mathematics education*. PedF UK.

Heyes, G., Sharmina, M., Mendoza, J. M. F., Gallego-Schmid, A., & Azapagic, A., (2018). Developing and implementing circular economy business models in serviceoriented technology companies. *Journal of Cleaner Production, 177*, 621–632.

Hutto, D. D., & Myin, E. (2013). *Radicalizing enactivism: Basic minds without content*. MIT Press.

Jones, K., & Sharp, M. (2007). A new performance-based process model for built asset maintenance. *Facilities, 25*(13/14), 525–535. https://doi.org/10.1108/02632770710822616

Juvova, A., Chudy, S., Neumeister, P., Plischke, J., & Kvintova, J. (2015). Reflection of constructivist theories in current educational practice. *Universal Journal of Educational Research, 3*(5), 345–349. https://doi.org/10.13189/ujer.2015.030506

Kayan, S. (2006). *Identity, culture, and learning in global contexts*. Routledge.

Khalid, E. I., Abdullah, S., Hanafi, M. H., Said, S. Y., & Hasim, M. S. (2019). The consideration of building maintenance at design stage in public buildings: The current scenario in Malaysia. *Facilities, 37*(13–14), 942–960. https://doi.org/10.1108/F-04-2018-0055

Kleidorfer, M., Möderl, M., Tscheikner-Gratl, F., Hammerer, M., Kinzel, H., & Rauch, W., (2013). Integrated planning of rehabilitation strategies for sewers. *Water Science & Technology, 68*, 176–183.

Koh, T. Y., & Low, S. P. (2010). Empiricist framework for TQM implementation in construction companies. *Journal of Management in Engineering, 26*(3), 133–143.

Lavy, S., Garcia, J. A., & Dixit, M. K. (2014). KPIs for facility's performance assessment, Part I: Identification and categorization of core indicators. *Facilities, 32*(5/6), 256–274. https://doi.org/10.1108/F-09-2012-0066

Leach, C. W., Van Zomeren, M., Zebel, S., Vliek, M. L. W., Pennekamp, S. F., Doosje, B., et al. (2008). Group-level self-definition and self-investment: A hierarchical model of in-group identification. *Journal of Personality and Social Psychology, 95*(1), 144–165. https://doi.org/10.1037/0022-3514.95.1.144

Ling, F. Y. Y., & Formoso, C., Wijeratne, W. M. P. U., Perera, B. A. K. S., & De Silva, L. (2014). Identification and assessment risks in maintenance operations. *Built Environment Project and Asset Management, 4*(4), 384–405.

Lu, W., Chen, K., Xue, F., & Pan, W. (2018). Searching for an optimal level of prefabrication in construction: An analytical framework. *Journal of Cleaner Production, 201*, 236–245.

Lückmann, P., & Färber, K. (2015). The impact of cultural differences on project stakeholder engagement: A review of case study research in international project management. *Procedia Computer Science, 100*, 85–94. https://doi.org/10.1016/j.procs.2016.09.127

MacGilchrist, B. (2003). Primary learners of the future: Laying the foundations for lifelong learning in the primary school. *Education, 3-13, 31*(1), 58.

Mahdavi, A. (2020). Explanatory stories of human perception and behavior in buildings. *Building and Environment, 168*, Article 106498. https://doi.org/10.1016/j.buildenv.2019.106498

Moradi, M., Abdollahzadeh, M. R., & Vakili, A. (2011). Effects of implementing 5S on total productive maintenance: A case in Iran. In *2011 IEEE International Conference on Quality and Reliability (ICQR)*, 14–17 September 2011.

Musa, M. F., Yusof, M. R., Mohammad, M. F., Mahbub, R., Alam, S., & Com, F., (2014). Characteristics of modular construction: Meeting the needs of sustainability and innovation. In *Colloquium on humanities, science and engineering innovative photovoltaic luminescent solar concentrator window: Energy and environmental aspects*; *Sustainability (Switzerland), 14*(7), 1–31. https://doi.org/10.3390/su14074292

Nannikar A. A., Raut, D. N., Chanmanwar, R. M., & Kamble, S. B. (2012). *FMEA for manufacturing and assembly process.*

Nsamenang, A. B. (2015). *Human development in cultural context: A third world perspective* (2nd ed.). Sage.

Ofori, I., Duodu, P. M., & Bonney, S. O. (2015). Establishing factors influencing building maintenance practices. *Journal of Economics and Sustainable Development, 6*(24), 184–193.

Østergård, T., Jensen, R. L., & Maagaard, S. E. (2016). Building simulations supporting decision making in early design—A review. *Renewable and Sustainable Energy Reviews, 61*, 187–201. https://doi.org/10.1016/j.rser.2016.03.045

Plant, K. L., & Stanton, N. A. (2012). The explanatory power of Schema Theory: Theoretical foundations and future applications in Ergonomics. *Ergonomics, 56*(1), 1–15. https://doi.org/10.1080/00140139.2012.736542

Pokrovskaia, N. N., Ababkova, M. Y., & Fedorov, A. A. (2019). Professional identity formation in the digital economy. *Education Sciences, 9*(3), 190. https://doi.org/10.3390/educsci9030190

Schnabel, T., Asendorpf, J. B., & Greenwald, A. G. (2013). Understanding and using the implicit association test. *European Review of Social Psychology, 24*(1), 1–50. https://doi.org/10.1080/10463283.2013.843966

Shaikh, P. H., Nor, N. B. M., Sahito, A. A., Nallagownden, P., Elamvazuthi, I., & Shaikh, M. (2017). Building energy for sustainable development in Malaysia: A review. *Renewable and Sustainable Energy Reviews, 75*, 1392–1403. https://doi.org/10.1016/j.rser.2016.11.128

Shohet, I. M., & Nobili, L. (2017). Application of key performance indicators for maintenance management of clinics facilities. *International Journal of Strategic Property Management, 21*(1), 58–71. https://doi.org/10.3846/1648715X.2016.1245684

Singh, K., & Ahuja, I. S. (2012). Justification of TQM–TPM implementations in manufacturing organisations using analytical hierarchy process: A decision-making approach under uncertainty. *International Journal Productivity and Quality Management, 10*(1), 69–84.

Skoda, J., & Doulik, P., et al. (2009). *Current issues of selected didactics* (p. 235). UJEP.

Syamilah, B. Y. (2006). *Maintenance management through strategic planning for public school in Malaysia.* Unpublished Master's Thesis, University Technology Malaysia.

Teravainen, V., Junnonen, J. M., & Ali-Loytty, S. (2018). Organizational culture: Case of the Finnish construction industry. *Construction Economics and Building, 18*(1), 48–69.

Tommelein, I. (2015). Journey toward lean construction: Pursuing a paradigm shift in the AEC industry. *Journal of Construction Engineering and Management, 141*(6), 04015005.

Trinh, M. T., & Feng, Y. (2020). Impact of project complexity on construction safety performance: Moderating role of resilient safety culture. *Journal of Construction Engineering and Management, 146*(2), 04019103.

Van Riel, W., Langeveld, J., Herder, P., & Clemens, F., (2014). Integrating road and sewer works: Risk attitude and costs. In *9th World Congress on Engineering Asset Management—WCEAM 2014*, 28–31 October, Pretoria, South Africa.

Van Kesteren, M. T. R., Rijpkema, M., Ruiter, D. J., Morris, R. G. M., & Fernández, G. (2014). Building on prior knowledge: Schema-dependent encoding processes relate to academic performance. *Journal of Cognitive Neuroscience, 26*(10), 2250–2261. https://doi.org/10.1162/jocn_a_00630

Van Riel, W., van Bueren, E., Langeveld, J., Herder, P., & Clemens, F. (2016). Decision-making for sewer asset management: Theory and practice. *Urban Water Journal, 13*(1), 57–68. https://doi.org/10.1080/1573062X.2015.1011667

Waghmare, S. N., Raut, D. N., Mahajan, S. K., & Bhamare, S. S. (2014). Failure mode effect analysis and total productive maintenance: A review. *International Journal of Innovative Research in Advanced Engineering (IJIRAE), 1*(6), 183–203.

Wiley, J. (2019). *The embodied mind and meaning.* Oxford University Press.

Wong, R. W. M., & Loo, B. P. Y. (2022). Sustainability implications of using precast concrete in construction: An in-depth project-level analysis spanning two decades. *Journal of Cleaner Production, 378*(September), Article 134486. https://doi.org/10.1016/j.jclepro.2022.134486

Xian Wei, C., & Syahrul Mat Radzuan, I. (2021). Building condition assessment and preventive maintenance of the Old Shophouses in Raub, Pahang. *Research in Management of Technology and Business, 1*(1), 1280–1293. http://publisher.uthm.edu.my/periodicals/index.php/rmtb

Yacob, S., Ali, A. S., & Peng, A. Y. C. (2016). Building condition assessment: Lessons learnt from pilot projects. *MATEC Web of Conferences, 66*, 00072.

Yu, X., Zhang, J., & Chen, Y. (2020). Identity construction in digital learning environments. *Computers & Education, 146*, 103750. https://doi.org/10.1016/j.compedu.2019.103750

Yun, S., & Choi, H. (2025). Circumventing rent controls with tenants' maintenance fees: Evidence from Korea. *Journal of Housing Economics, 67*, Article 102042. https://doi.org/10.1016/j.jhe.2025.102042

Chapter 5
Global Perspective of Modularity Methodology in the Construction Industry

Abstract This chapter presents a global perspective on modularity methodology in the construction industry, offering insights into its guiding principles, growth drivers, economic implications, and application alignment. It begins by outlining the fundamental principles of modular construction methodologies, including standardisation, componentisation, flexibility, and integration of quality control. These principles serve as the foundation for efficient design, manufacturing, and assembly processes, enabling consistent quality and reduced construction timelines. The chapter then examines the factors influencing the growth of modularity methodologies worldwide. These include advancements in technologies, rapid urbanisation, labour shortages in traditional construction, and the need for large-scale infrastructure projects. An economic analysis follows, assessing the cost–benefit aspects of modular construction. The discussion considers initial capital investment, operational efficiency, lifecycle cost savings, and return on investment. This analysis highlights that while upfront costs may be higher, long-term economic and environmental gains make modularity an attractive option for many stakeholders. Finally, the chapter addresses the importance of matching modularity methodology applications to specific phenomena within the built environment. It explores how factors such as project scale, function, location, and cultural context influence the choice and success of modular systems. Overall, the chapter underscores that modularity methodology, when applied strategically and contextually, holds significant potential to transform the global construction industry, delivering faster, more sustainable, and cost-effective building solutions.

5.1 Introduction

Through the lens of institutional perspectives, which are based on organisational and sociological theories, the global perspective of modularity methodology in the construction industry can be comprehended effectively. These viewpoints highlight how organisational behaviour, industry practices, and the adoption of technological innovations are influenced by norms, values, beliefs, and regulatory frameworks.

P. Adekunle et al., *Modularity Methodology*,
https://doi.org/10.1007/978-3-032-12918-5_5

Institutional perspectives on modular construction shed light on how institutional forces influence the uptake, spread, and development of modular techniques in various markets and geographical areas. The idea of institutional isomorphism, which describes how organisations converge towards comparable structures, practices, and norms as a result of pressures from their institutional environments, is fundamental to institutional perspectives. According to this phenomenon, the adoption and normalisation of modular construction methods are greatly influenced by elements like industry standards, legal requirements, and professional norms. To legitimise modularity as a preferred construction method and to guide practice, industry organisations, regulatory agencies, and professional associations, for example, frequently create norms and standards.

Path dependence and historical legacies, which can both support and impede innovation, are also highlighted by institutional perspectives (Gann et al., 2000). The transition to modular solutions may be slowed in areas with deeply ingrained traditional construction methods due to vested interests and established norms. Though at varying rates across different industry segments, change is being progressively sparked by increased institutional pressure for innovation and growing awareness of the advantages of modularity, such as increased sustainability, efficiency, and quality. These viewpoints also emphasise how crucial coalitions and actor networks are to advancing technological change (Dubois & Gadde, 2002). To promote modular construction through policy advocacy, capacity building, and knowledge sharing, key actors, from government agencies and industry associations to technology providers and advocacy groups, can establish strategic alliances. These networks have the capacity to establish institutional environments that are conducive to the integration of modularity into popular construction methods by mobilising resources, forming partnerships, and influencing narratives surrounding modularity.

In general, the application of institutional perspectives to the modularity methodology demonstrates that institutional alignment is just as important to its worldwide adoption as technological viability. Together, these elements, regulatory backing, industry cooperation, and cultural acceptance, determine the rate and scope of modularity's adoption in the global construction sector. The underlying principles, growth drivers, economic implications, and real-world applications of modular construction methodologies will be examined in later sections of this chapter, which are set up by this foundation.

5.2 Principles of Modularity Construction Methodologies

Modularity methodology is a beacon of innovation within the construction industry, heralding a paradigm shift in how projects are conceptualised, designed, and executed (Munaro et al., 2020). The principles of modularity methodology offer a systematic framework for optimising construction processes, enhancing project efficiency, and fostering adaptability in an ever-evolving built environment. Exploration of the foundational principles that underpin modularity methodology in construction is vital in unravelling its essence, elucidating its significance, and examining its implications for the future of the construction industry. As underscored by Dube et al. (2014), modularity is a design principle that fosters standardisation, specialisation, and integration across diverse domains, enabling construction practitioners to achieve economies of scale, mitigate risks, and accelerate innovation in project delivery. Construction stakeholders can then navigate the complexities of modern construction projects with greater agility, efficiency, and foresight, thereby unlocking new opportunities for collaboration, creativity, and value creation in the built environment. This review illuminates the underlying mechanisms and emergent patterns that characterise the application of modularity methodology in construction projects worldwide through a synthesis of theoretical insights, empirical evidence, and practical examples. A deeper understanding of the principles of modularity methodology will inspire critical reflection, informed dialogue, and actionable insights that can empower construction stakeholders to embrace modularity as a catalyst for innovation, resilience, and sustainable development in the 21st-century built environment.

5.2.1 Standardisation

Standardisation offers a systematic approach to designing, manufacturing, and assembling modular components within the built environment (O'Connor et al., 2015). Standardisation entails the establishment of uniform specifications, dimensions, and interfaces for modular elements, enabling interoperability, interchangeability, and compatibility across diverse construction projects. Modularity methodology seeks to streamline construction processes, enhance project efficiency, and promote economies of scale by embracing standardisation. This will drive innovation, reduce costs, and accelerate project delivery timelines. Pervez et al. (2022) identified that one of the primary goals of standardisation within modularity methodology is to foster consistency and repeatability in construction practices, ensuring that modular components adhere to predefined guidelines, codes, and quality standards. Construction stakeholders can use standardised dimensions, materials, and manufacturing processes to minimise variations, errors, and inconsistencies in component fabrication, assembly, and installation, mitigating risks and improving project predictability. As seen in Fig. 5.1, O'Connor et al. (2015) suggested two approaches to integrating design standardisation with modularisation. One approach is to standardise plant designs with no modularisation, and the other is to modularise plants with no design standardisation. Companies can vary, using one or both strategies to accomplish design standardisation and modularisation.

Standardisation catalyses innovation and knowledge transfer within the construction industry, enabling the development of modular solutions that can be replicated, adapted, and scaled across different projects and contexts (Featherston et al., 2016). Establishing standard design parameters and performance criteria through standardisation facilitates collaboration, interoperability, and cross-pollination of

Fig. 5.1 Two approaches to integrating design standardisation. *Source* O'Connor et al. (2015)

ideas among architects, engineers, manufacturers, and construction professionals, fostering a culture of continuous improvement and collective learning within the industry. Standardisation is pivotal in driving down costs and enhancing cost-effectiveness in construction projects (Hong et al., 2018). Standardised modular components can be mass-produced, procured, and deployed at lower costs than bespoke, site-built alternatives through economies of scale and scope. Using standardised components, construction stakeholders can optimise resource allocation, minimise waste, and achieve greater cost efficiency throughout the project lifecycle, enhancing overall project affordability and competitiveness in the marketplace. Aside from its economic benefits, standardisation enhances safety, quality, and regulatory compliance in construction practices. Sharma and Chandel (2013) argued that adhering to established industry standards, modular components undergo rigorous testing, certification, and quality assurance processes to ensure structural integrity, performance, and durability. These standardised components facilitate compliance with building codes, zoning regulations, and environmental requirements, enhancing project resilience and sustainability in the face of evolving regulatory landscapes and market dynamics.

5.2.2 *Componentisation*

Componentisation emphasises the systematic decomposition of complex systems into discrete, interchangeable components (Jaskolka et al., 2020). This principle seeks to enhance flexibility, scalability, and efficiency in construction projects by breaking down structures, processes, and workflows into modular units that can be designed, manufactured, and assembled independently and integrated seamlessly to form cohesive systems and structures. Componentisation in construction involves identifying, standardising, and modularising building elements such as walls, floors, ceilings, mechanical systems, and architectural features (Bender et al., 2015). Each component is designed to fulfil specific functions and requirements while maintaining standardised interfaces and compatibility with other components, facilitating ease of assembly, maintenance, and future modifications.

The principle of modularity methodology is expressed in componentisation, which promotes design flexibility and customisation while ensuring consistency and reliability in construction practices, as defined by Dajsuren et al. (2013). Breaking down complex structures into modular components, architects and designers gain the flexibility to mix and match components to meet diverse project needs, accommodate site-specific constraints, and respond to changing user requirements. The standardised interfaces and design guidelines enable designers to create modular components which are easily integrated into different architectural styles, building typologies, and construction contexts (Stephan et al., 2013). This promotes design innovation and creative expression in the built environment. In addition, componentisation facilitates the standardisation and optimisation of construction processes, enabling the mass production, prefabrication, and assembly of modular components in controlled factory environments. By standardising dimensions, materials, and manufacturing techniques, construction stakeholders can streamline production workflows, minimise waste, and enhance quality control throughout manufacturing and assembly (Schwinghammer & Henseler, 2010). This will aid the prefabrication of modular components manufactured off site and transported to the construction site for rapid assembly, thereby reducing construction time, labour costs, and on-site disruptions.

Another critical principle of componentisation is to enhance project scalability and adaptability, allowing construction projects to evolve and expand over time in response to changing user needs, market conditions, and technological advancements (MacCormack et al., 2007). By adopting a modular approach to construction, developers and owners can incrementally add or replace modular components as needed without disrupting the overall structure or functionality of the building. This flexibility enables buildings to adapt to evolving usage patterns, accommodate future expansions, and incorporate emerging technologies, enhancing their longevity, value, and sustainability over the project lifecycle. Also, componentisation fosters interoperability and compatibility among different building systems and components (Rau, 2001) thereby enabling seamless integration of architectural, structural, mechanical, and electrical elements within the built environment. By adhering to standardised interfaces, protocols, and communication protocols, modular components can interact and communicate with each other, creating intelligent, interconnected buildings that optimise energy efficiency, occupant comfort, and operational performance.

5.2.3 *Flexibility and Adaptability*

Flexibility and adaptability are aimed to enhance the resilience, responsiveness, and longevity of built environments in the face of dynamic and evolving requirements, contexts, and challenges (Khoder & Tommelein, 2017). These principles emphasise the ability of modular construction systems and components to accommodate changing needs, preferences, and conditions over time, thereby enabling construction projects to remain relevant, functional, and sustainable throughout their lifecycle. Flexibility in modular construction refers to the inherent capacity of modular systems and components to be easily modified, reconfigured, or repurposed to meet diverse project requirements, site constraints, and user preferences (Vasawade et al., (2015). Modular construction facilitates design flexibility by breaking down complex structures into interchangeable components assembled, disassembled, and recombined in various configurations to adapt to evolving needs and contexts. This flexibility enables architects, designers, and developers to respond quickly and efficiently to changing market demands, regulatory requirements, and technological advancements without necessitating extensive redesign or reconstruction.

Adaptability, on the other hand, refers to the ability of modular construction systems and components to respond proactively to changes in external conditions, such as environmental, economic, and social factors, while maintaining their functionality, performance, and relevance (Hou & Qian, 2016). Modular construction enables buildings to adapt to changing environmental conditions, such as climate change, extreme weather events, and resource scarcity, by incorporating sustainable design principles, energy-efficient technologies, and resilient building materials that enhance their durability, efficiency, and resilience in the face of environmental challenges. Furthermore, adaptability in modular construction enables buildings to evolve and grow over time in response to changing user needs, market dynamics, and technological innovations. Modular buildings can be designed with modular expansion capabilities, allowing for incremental additions, modifications, and upgrades to accommodate future growth, expansions, or changes in building usage. This adaptability enables buildings to remain viable and competitive in the long term, mitigating the risk of obsolescence and ensuring their continued relevance and value in a rapidly changing world.

Martin et al. (2016) revealed that flexibility and adaptability emerge as essential principles and objectives of modularity methodology in construction, enabling buildings to remain responsive, resilient, and sustainable in the face of evolving needs, contexts, and challenges. Modular construction offers a versatile and future-proof approach to building design and construction through flexibility and adaptability. It enables buildings to evolve and thrive in response to changing conditions, user preferences, and market dynamics. As the construction industry embraces modular approaches, flexibility and adaptability will remain guiding principles that drive innovation, resilience, and sustainability in the built environment.

5.2.4 Efficiency and Timesaving

The modularity methodology in construction emphasises efficiency and timesaving as fundamental concepts. The ultimate goal of this approach is to optimise resource utilisation, simplify procedures, and expedite project delivery deadlines (Innella et al., 2019). These principles emphasise the systematic organisation, standardisation, and modularisation of construction components and processes to enhance productivity, reduce waste, and minimise project duration. In modularity methodology, efficiency maximises output while minimising inputs, thus optimising resource utilisation and project performance. Breaking down complex construction processes into modular components permits construction stakeholders to identify and eliminate inefficiencies, redundancies, and bottlenecks that often arise in traditional construction practices (Naim & Baarlow, 2010).

Modular construction further enables the standardisation of materials, components, and workflows, facilitating economies of scale and scope that drive down costs, improve quality, and enhance overall project efficiency. Notably, Popescu (2016) argued that efficiency in modular construction extends beyond cost savings to encompass time-saving benefits that stem from the off-site fabrication, prefabrication, and assembly of modular components in controlled factory environments. Manufacturing building components concurrently with site preparation and foundation works enables parallel processing of construction activities in modular construction. Consequently, this leads to the compression of project schedules and reduces overall construction duration. This time-saving advantage accelerates project delivery and minimises on-site disruptions, safety risks, and environmental impacts associated with traditional construction methods.

Modularity methodology emphasises the importance of standardisation and repetition in achieving efficiency gains in construction projects

. This will aid modular components' production, prefabrication and assemblage with incredible speed and precision, minimising errors, rework, and delays commonly encountered in site-built construction. The repetitive nature of modular construction enables construction teams to develop specialised expertise, optimise workflows, and implement lean manufacturing principles that enhance productivity and quality across multiple projects (Love et al., 2003). Therefore, in enhancing efficiency, modularity methodology can prioritise time-saving benefits that result from the systematic organisation and coordination of construction activities through modularisation. By breaking down projects into discrete, manageable modules, construction teams can plan, schedule, and execute tasks more precisely, minimising idle time, downtime, and rework while maximising productivity and throughput (Moghadam et al., 2012). Modular construction enables the parallelisation of construction activities, allowing multiple modules to be fabricated, transported, and installed simultaneously, thereby shortening project duration and accelerating time to market.

5.2.5 Cost-Effectiveness

Cost-effectiveness is a fundamental principle and a central objective of modularity methodology in construction, aiming to optimise resource allocation, minimise waste, and enhance value creation throughout the project lifecycle (Quale et al., 2012). This principle emphasises the systematic organisation, standardisation, and modularisation of construction components and processes to achieve the greatest possible return on investment while meeting project requirements and stakeholder expectations. Cantu et al. (2019) defined the scope of cost-effectiveness within modularity methodology as reducing overall project costs while maintaining or improving quality, functionality, and performance. Modular construction enables cost savings through various means, including economies of scale, streamlined workflows, and reduced material waste. Modular components in construction can be mass-produced, prefabricated, and assembled with incredible speed and precision, thereby minimising labour costs, material costs, and overhead expenses associated with traditional construction methods. Cost-effectiveness in modular construction extends beyond initial construction costs to encompass long-term operational and lifecycle costs, as observed by Olawumi et al. (2022). Modular buildings are designed with sustainability and energy efficiency in mind, incorporating green building materials, energy-efficient systems, and passive design strategies that reduce energy consumption, lower operating costs, and minimise environmental impact over the building's lifespan. Modular buildings can achieve significant cost savings in utilities, maintenance, and lifecycle management by optimising energy performance and enhancing overall cost-effectiveness and return on investment for building owners and occupants.

Furthermore, modularity methodology emphasises the importance of value engineering and cost optimisation in achieving cost-effectiveness in construction projects (Noordzy et al., 2021). By breaking projects into modular components, construction stakeholders can identify cost-saving opportunities, optimise design decisions, and prioritise investments based on their potential to deliver maximum value and performance. The repetitive nature of modular construction also enables construction teams to refine processes, implement lean manufacturing principles, and leverage technological innovations that enhance productivity, quality, and cost-effectiveness across multiple projects (Blismas & Wakefield., 2009). In addition to reducing costs, modularity methodology emphasises the importance of enhancing value creation and stakeholder satisfaction in construction projects. This is achieved by standardising components, streamlining workflows, and minimising construction timeframes. Modular construction also enables stakeholders to realise faster project delivery, earlier occupancy, and quicker returns on investment, enhancing overall project viability, competitiveness, and financial performance, as Hsu et al. (2018) observed. Subsequently, this ensures greater predictability and reliability regarding cost estimates, schedules, and project outcomes, reducing financial risks and uncertainties associated with traditional construction methods.

5.2.6 Quality Control

Quality control ensures that modular components and systems meet predefined standards, specifications, and performance criteria throughout the project lifecycle (Rausch et al., 2020). This emphasises systematically identifying, preventing, and resolving defects, errors, and deficiencies in modular construction processes and products to deliver high=quality, reliable, and durable buildings that meet or exceed stakeholder expectations. Quality control within modularity methodology is to minimise variations, deviations, and inconsistencies in construction processes and products through rigorous inspection, testing, and verification procedures (Enshassi et al., (2020). Modular construction enables quality control by standardising dimensions, materials, and manufacturing techniques, ensuring uniformity, consistency, and reliability across modular components and systems. Following established quality standards, construction stakeholders can identify and address potential quality issues early in the construction process, thereby minimising rework, delays, and cost overruns associated with poor-quality construction practices.

Bertram et al. (2019) observed that quality control in modular construction extends beyond the manufacturing phase to encompass all stages of the project lifecycle, including design, procurement, transportation, assembly, and installation. Modular construction enables quality control by implementing robust quality management systems, performance metrics, and feedback mechanisms such as the waterfall model, as seen in Fig. 5.2. These enable construction stakeholders to monitor, evaluate, and improve the quality of modular components and systems throughout the project lifecycle. Establishing clear quality objectives, responsibilities, and accountability mechanisms ensures that quality remains a top priority throughout the project lifecycle. These aid construction stakeholders in enhancing customer satisfaction, stakeholder confidence, and project success. Modularity methodology emphasises the importance of continuous improvement and innovation in achieving quality excellence in construction projects (Egege, 2018). By leveraging data analytics, feedback loops, and lessons learned from previous projects, construction stakeholders can identify trends, patterns, and areas for improvement in modular construction processes and products, thereby driving innovation, efficiency, and quality improvement over time. Modularity methodology also encourages collaboration, knowledge sharing, and best practices dissemination among construction stakeholders, fostering a culture of quality excellence and continuous improvement within the construction industry (Guerra & Leite, 2021). In addition to enhancing product quality, modularity methodology emphasises the importance of ensuring modular buildings' safety, durability, and performance under various environmental conditions, usage scenarios, and operational contexts. Modular construction further enables quality control by incorporating robust design principles, structural analysis, and testing protocols (Baduge et al., 2022). This ensures that modular buildings can withstand various hazards, loads, and stresses while maintaining structural integrity, occupant safety, and building performance over time. By prioritising rigorous quality standards and performance criteria, modular buildings can achieve certification, accreditation, and compliance with regulatory requirements, thereby enhancing stakeholder confidence and trust in the quality and reliability of modular construction practices.

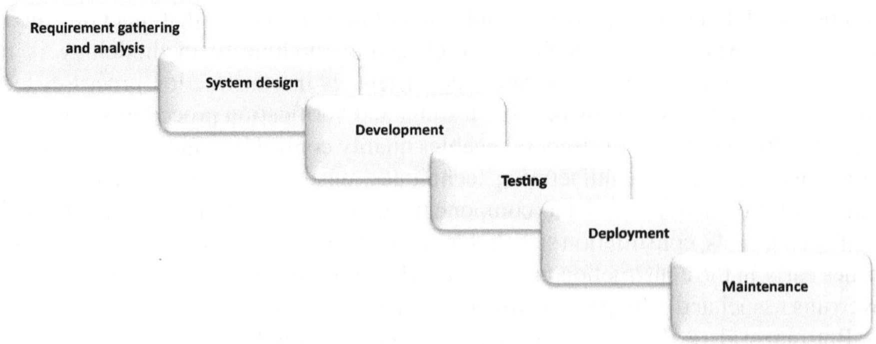

Fig. 5.2 Waterfall model. *Source* Prasetya et al. (2021)

5.2.7 Scalability and Replicability

Scalability and replicability mean facilitating the flexible expansion, replication, and adaptation of construction projects to meet diverse needs, contexts, and requirements. This principle emphasises the systematic organisation, standardisation, and modularisation of construction components and systems to enable projects to scale up or down in size, complexity, and scope while maintaining consistency, efficiency, and quality across multiple iterations. Scalability in modularity methodology also enables construction projects to adapt and grow in response to changing needs, market conditions, and stakeholder requirements (Bjornfot & Stehn, 2007). Modular construction facilitates scalability by breaking down projects into discrete, interchangeable modules that can be easily replicated, modified, or expanded to accommodate changes in building usage, occupancy levels, and functional requirements.

Moreover, scalability in modular construction extends beyond physical expansion to encompass scalability in design, technology, and innovation. Modular construction enables designers and developers to create flexible, adaptable building systems and components that can be customised, modified, and reconfigured to meet diverse aesthetic, functional, and performance requirements. Construction stakeholders leveraging modular design principles can incorporate scalable technologies, such as intelligent sensors, IoT devices, and renewable energy systems that can be easily integrated into modular buildings to enhance their functionality, efficiency, and sustainability over time.

Modularity methodology emphasises the importance of replicability in achieving scalability and calibration in construction projects (Khalfan & Maqsood, 2014). Standardising components, processes, and workflows will enable construction stakeholders to replicate successful modular solutions across different projects, locations, and contexts. Project efficiency, consistency, and quality will also be enhanced by leveraging economies of scale, knowledge sharing, and dissemination of best practices (Lessing & Stehn, 2019). This implies that replicability enables construction stakeholders to capitalise on lessons learned, optimise workflows, and refine processes that drive continuous improvement and innovation in modular construction practices. In addition to enhancing scalability and replicability, Rocha and Koskela (2020) revealed that modularity methodology promotes interoperability and compatibility among different building systems and components, enabling seamless integration and interaction between modular components and systems within the built environment. With protocols and communication standards in place, modular components can interact and communicate with each other, creating interconnected, interoperable buildings that optimise energy efficiency, occupant comfort, and operational performance.

5.2.8 Safety and Resilience

Safety and resilience protect life, property, and the environment while enhancing built structures' durability, reliability, and adaptability in the face of various hazards, risks, and uncertainties (Shahtaheri et al., 2017). Modularity methodology is vital to minimise risks, hazards, and accidents associated with construction processes, materials, and operations. Modular construction promotes safety by reducing on-site construction activities, minimising exposure to hazardous conditions, and mitigating the risk of accidents, injuries, and fatalities (Haas et al., 2000). Modular construction eliminates many safety hazards and risks associated with traditional on-site construction, such as falls, collapses, and exposure to hazardous materials by shifting construction activities to controlled factory environments. This will ensure safer working conditions for construction workers and reduce the likelihood of accidents and injuries. Song et al. (2005) argued that safety in modular construction extends beyond the construction phase to encompass the entire building lifecycle, including occupancy, maintenance, and decommissioning. Modular buildings are designed with safety in mind, incorporating robust structural systems, fire-resistant materials, and advanced safety features that enhance occupant safety, security, and well-being.

Modular construction ensures that buildings are structurally sound, fire-resistant, and compliant with safety codes and regulations by adhering to stringent safety standards and regulatory requirements (Nahangi et al., 2015). This will reduce the risk of accidents, injuries, and liabilities associated with building operations and occupancy. Furthermore, modularity methodology emphasises the importance of resilience in ensuring the long-term durability, reliability, and performance of built structures under various environmental, operational, and societal conditions (Lu & Korman, 2010). Modular construction enables resilience by incorporating redundant systems, flexible design features, and adaptive technologies that enhance the ability of buildings to withstand and recover from adverse events, such as natural disasters, climate change, and technological disruptions. Construction stakeholders can then create buildings capable of adapting to changing conditions, minimising downtime, and maintaining functionality and usability in the face of disruptions and disturbances by integrating resilient design principles. Simpeh et al. (2015) also suggest that modularity methodology promotes sustainability and environmental stewardship in construction practices and enhances safety and resilience. Modular construction enables sustainability by optimising resource utilisation, minimising waste, and reducing environmental impact throughout the project lifecycle. The method minimises material waste, energy consumption, and environmental pollution associated with traditional construction methods, reducing construction projects' carbon and ecological footprint.

5.2.9 Energy Efficiency and Optimisation

Reflecting the industry's increasing focus on sustainability, resource conservation, and environmental stewardship, energy efficiency and optimisation represent fundamental principles of modularity methodology in construction. As the construction sector confronts the challenges of climate change, resource scarcity, and energy consumption, integrating energy-efficient practices and optimisation strategies becomes paramount in enhancing the sustainability and resilience of built environments. Modularity methodology in construction emphasises designing, fabricating, and assembling engineered modular components for optimal energy performance. Through advanced materials, building technologies, and energy-efficient systems, modular construction offers opportunities to minimise energy consumption, reduce greenhouse gas emissions, and mitigate the environmental footprint of construction projects. Innovative design solutions, such as passive solar design, natural ventilation, and high-performance insulation, can enable modular buildings to achieve superior energy efficiency and thermal comfort while reducing reliance on mechanical heating, cooling, and lighting systems.

Ibn-Mohammed et al. (2013) postulated that modularity methodology enables implementation of integrated energy management systems and intelligent building technologies that optimise energy usage, monitor performance metrics, and enhance operational efficiency throughout the building lifecycle. Building owners and operators can monitor energy consumption patterns and identify optimisation opportunities in modular buildings by incorporating sensors, controls, and automation features. Adjusting building settings in real-time to maximise energy savings and minimise environmental impact will also be possible. As part of the objectives, modular construction allows for the seamless integration of renewable energy technologies, such as solar panels, wind turbines, and geothermal systems, which further enhance the energy efficiency and sustainability of buildings by harnessing clean, renewable energy sources. Modularity methodology also facilitates off-site fabrication and precision manufacturing processes that minimise material waste, optimise resource utilisation, and streamline construction workflows, thereby reducing construction projects' embodied energy and environmental footprint.

The prefabrication of modular components in controlled factory settings leads to lower environmental impact and improved sustainability outcomes. Construction waste and transportation-related emissions are also significantly reduced.

Through energy optimisation, modular construction methods enable faster project delivery times and shorter construction schedules. This further contributes to energy savings and environmental conservation by reducing onsite energy consumption and construction-related disturbances. Modular construction provides a technique to achieve net-zero energy buildings, lower carbon emissions, and encourages environmental sustainability in the construction sector by combining energy-efficient design ideas, innovative technology, and renewable energy sources.

5.2.10 Maximum Utilisation of Resources

Reflecting the industry's growing awareness of the environmental impacts associated with material extraction, processing, transportation, and disposal has aided the maximum utilisation of resources (Kamali et al., 2019). As concerns about resource scarcity, embodied energy, and carbon emissions continue to mount, adopting modular construction methods offers opportunities to optimise material usage, minimise waste, and enhance the sustainability of building projects. Modularity methodology prioritises efficiently utilising materials through off-site fabrication and prefabrication processes. Construction firms can optimise material usage by minimising over-ordering, reducing cutting and trimming waste, and optimising production processes to maximise material yield through modularisation. Through digital modelling, parametric design, and computer-aided manufacturing (CAM) technologies, modular construction enables precise material optimisation and efficient use of resources, resulting in lower material consumption and reduced environmental impact.

Modularity methodology also facilitates the implementation of sustainable material choices and alternative construction methods that prioritise energy efficiency, durability, and recyclability. This is achieved by selecting environmentally-friendly materials, such as recycled content, low-impact materials, and renewable resources. Modularity in construction projects can minimise embodied energy and carbon emissions associated with material production and manufacturing. Modular construction incorporates modular systems, such as modular façade panels, modular flooring systems, and modular wall assemblies. These systems can be reused, repurposed, or recycled at the end of their lifecycle, enhancing resource efficiency and circularity. Supply chain optimisation and logistics management practices that reduce energy consumption and greenhouse gas emissions associated with material transportation and delivery are promoted in modularity practices. Centralising manufacturing operations in dedicated factory facilities near transportation hubs or project sites will enable modular construction firms to minimise transportation distances, optimise freight routes, and consolidate material shipments, thereby reducing fuel consumption, vehicle emissions, and traffic congestion. This construction method will also enable just-in-time delivery and lean construction practices that minimise on-site material storage, handling, and waste, leading to greater operational efficiency and environmental sustainability. Modular construction offers a pathway towards achieving greater energy efficiency, material optimisation, and environmental sustainability in building projects. Stakeholders in the built environment can leverage the transformative potential of modularity methodology to create more efficient, resilient, and environmentally responsible built environments for current and future generations.

5.3 Factors Influencing the Growth of Modular Construction

There has been a growing demand for modular construction globally in recent years. Modular construction is a process whereby a building is constructed off site using the same materials and designed to the same standards as conventionally built facilities—but in about half the time. Several technological advancements have made modularity methodology in construction more accessible, such as BIM and the development of sophisticated computer-aided design (CAD) systems. On the other hand, the skilled labour shortage and the desire for eco-friendly, sustainable construction solutions have prompted many construction professionals to explore modular construction (Feldmann et al., 2022). In addition, the increasing rate of urbanisation has encouraged the adoption of modular construction practices. With the unprecedented growth in global infrastructure projects, urban developers and certifiers have started to recognise the value of modular construction owing to its ability to offer a much-needed flexible solution in large-scale and smaller urban projects. Several factors beyond technology have contributed to the rise of modular construction. For example, the potential for mass customisation due to changes in consumer preference can lead to the development of more innovative modular construction practices. Explained below are the significant factors influencing the growth of modular construction globally.

5.3.1 Technological Advancement

Over the past few decades, significant technological advancements have transformed the global construction industry (Pan & Zhang, (2023). Modular construction has particularly benefitted from these technological innovations. Using advanced management software to streamline the design and production process has improved construction accuracy and efficiency. BIM uses computer-generated 3D models to visualise a building and simulate its construction (Wuni & Shen, (2020). When architects, engineers, and construction professionals plan a project using BIM, they can generate a 3D model that includes every aspect of the building process. As a result, BIM allows for greater precision in design, more efficient use of materials, reduced waste, and more effective collaboration between different teams. These translate into faster and more efficient construction, a significant selling point of modular construction. Moreover, BIM could enable the creation of 'smart buildings', where sensors and internet-enabled devices can be installed (Razkenari et al., 2020). These devices can connect and communicate with each other over the Internet and be remotely monitored and controlled. The creation of such intelligent modular buildings has the potential to reduce ongoing construction costs significantly, facilitate more effective building management and maintenance, and improve the living and working environment for building users.

Furthermore, Abdelmageed and Zayed's (2020) technological advancement in modular construction has also led to increasing modular capacity as improved manufacturing processes and sophisticated machinery result in higher production and greater quality control and flexibility in design. The modular industry is estimated to potentially deliver significantly more significant amounts of new homes and commercial buildings. For example, the 'Get It Right' report from the Modular and Portable Building Association (MPBA) estimates that modern methods of construction, which include modular construction, have the potential to deliver between 100,000 and 150,000 new homes each year (Salama et al., 2021).

In addition to reducing the time taken to construct a building, technological advancements have also directly contributed to improving the product life and quality of modular buildings, a notion reflected in UK building regulations and standards. For example, the British Board of Agreement (BBA) is a leading construction certification body that provides manufacturers with approval and an independent and comprehensive evaluation of modular buildings' physical, mechanical, chemical, and functional properties. With the aid of advanced technology, modular construction methods are now well-positioned to meet the requirements of such standards. Given the recent stringency of UK regulations and its zero-carbon environmental target for all new homes from 2016, to achieve the optimum sustainability level and to ensure a long building life, modular construction technologies, with the capability to reflect the most up-to-date sustainable and environmentally friendly materials and processes—might prove to be ever more sought after as a modern construction method. This suggests that technological developments provide construction solutions and offer consumers a potentially more sustainable and long-term building option.

5.3.2 Skilled Labour Shortage

Ghannad et al. (2020) observed that workers in modular construction are sometimes expected to have higher skill levels than workers in traditional construction because they are responsible for completing more detailed and complex work in isolated modules. However, this assumption has been widely discredited, with many studies suggesting that modular construction has the same—if not fewer—skills-based demands than traditional construction. The Urban Land Institute's report "Permanent Modular Construction 2015" (p. 19), found that "the relatively low skill level required for that portion of the process lessens the significance of labour availability for modular construction, thereby reducing the impact of a labour shortage". However, while the requirement for skilled labour in permanent modular construction may not hurt the sector when the construction industry sees substantial workforce challenges, studies have shown that traditional construction suffers from a skills deficit.

This may be another factor driving the growth of modular construction—the ongoing difficulty in finding traditional skills may push more investors and companies towards exploring other construction methods. Further to the expert opinion that the skills required in modular and traditional construction do not differ significantly, it is worth noting that students and schools have yet to adopt modular construction widely into curriculums and qualifications. Vocational skills for traditional construction remain the focus of many further educational institutions, meaning that the skills gap limitation for traditional construction perpetuates itself.

It appears that widespread misconceptions and a lack of educational headway prevent the merits and strengths of modular construction in addressing worker shortages from being realised in practice. In a construction environment already experiencing the impact of the skills shortage and an uncertain post-Brexit UK, it seems that the potential of modular construction to help plug the skills gap may be a significant benefit for both the construction industry and the wider community. Jeong et al. (2022) posit that increasing skilled immigration and improving education programmes for potential construction workers should be strategies explored by the sector. However, these changes inevitably require time, capital, and government support to have a positive outcome. Given the rise of modular construction developments, it is worth considering how we might harness the modular methods' potential to reduce reliance on traditional skills and push towards more research and better practices within this specialist construction area. After all, successfully implementing and spreading modular construction can help address the ongoing skills shortage in construction, along with the other well-documented benefits of this innovative process.

5.3.3 Rapid Urbanisation

Urbanisation is the process by which people migrate from rural to urban areas, increasing the proportion of people living in cities (Pan & Zhang, 2023). This global phenomenon has been a significant driving force in the growth of modular construction. Rapid urbanisation presents many challenges regarding traditional construction methods, particularly in responding to the enormous demand for new spaces and maintaining minimal disruption in densely populated areas. Modular construction solves these challenges, enabling parallel design and site work processing. Off-site production can happen simultaneously with foundation and other preparatory work. Modular construction can provide faster, more efficient building solutions in urban areas by maximising factory production and minimising the impact on the local community. This is essential when governments worldwide are increasingly focused on developing smart cities, using data and advanced technologies to manage the growing demand for new services and facilities. Smart cities require real-time data about the urban environment and infrastructure, which can be captured, managed and analysed more comprehensively using intelligent modular construction techniques (Dipasquale et al., 2020).

This link between the growth of modular construction and the development of intelligent cities represents a mutually beneficial cycle. As observed in Fig. 5.3, Zhang et al. (2022, p. 6) stated that "the influencing factors of the urban innovation and development of cities are environmental quality, geographical location, and city scale from a natural perspective; economic level, industrial structure, industrial agglomeration, and technological innovation from an economic perspective; and administrative hierarchy, cultural environment, "population structure, government management, and service measures from a social perspective". On the one hand, modular construction offers a method for quickly incorporating innovation and technological advancement into urban building projects. On the other hand, the drive towards more innovative, data-driven cities provides an external push for the rapid growth and further development of modular construction solutions. Population growth is another factor that has increased owing to rapid urbanisation, further boosting the demand for modular construction. Modular construction can offer a quicker, more affordable alternative because complicated, large-scale building projects frequently need time-consuming, expensive, and less flexible traditional construction methods. It also allows developers to increase the pace of construction work, adapting more flexibly to the changing needs of urban communities and evolving infrastructural requirements. As major cities are identifying and embracing the opportunities for large-scale modernisation of transport, health, education and housing provisions, the conditions are ripe for the continued growth of modular construction in the foreseeable future.

5.4 Maximum Utilisation of Resources

5.4.1 Maximum Utilisation of Resources

5.4.2 Large-Scale Infrastructure Projects

For many years, the image of modular systems has been limited to small, single-storey buildings (Jiang et al., (Jiang 2022). However, technological advancements over the past 10–15 years, in particular, have helped to facilitate off-site modular construction and made it an increasingly viable and attractive option for large-scale, multi-storey projects. The ability to produce 3D volumetric modules has been a critical factor, as it enables an entire room or apartment to be assembled in one single modular unit in the factory, meaning that the finished buildings no longer have to show the lines and fixing

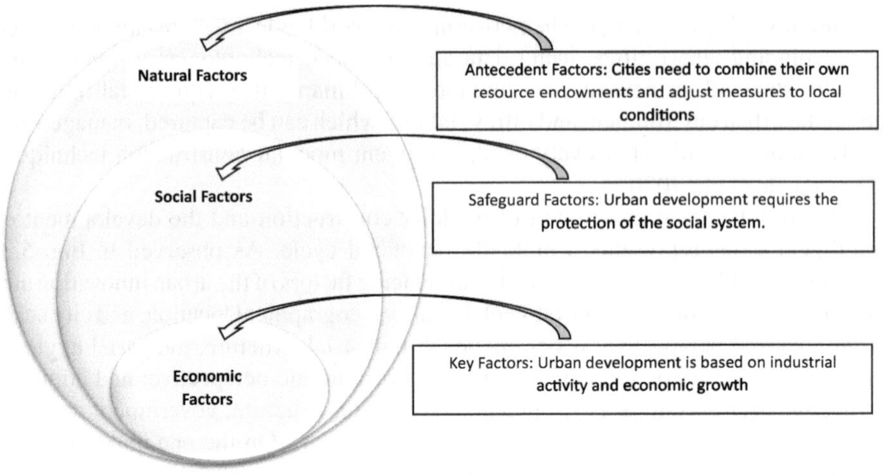

Fig. 5.3 Onion model of factors affecting urban innovation and development. *Source* Zhang et al. (2022)

points of individual modules. Instead, once assembled, the modules are indistinguishable from the traditionally constructed buildings. The construction sector can begin to change the frequently unfavourable impression of the sector, which is characterised by delays, disruptions, and a lack of skilled workers, by embracing the opportunities that modular construction presents for large-scale projects. Throughout the UK and internationally, we have seen the successful implementation of large-scale modular construction projects, which has led to a shift in many common assumptions about modular construction and increased interest and investment in the process. Large-scale projects that place confidence in the collaboration between designers, contractors, and manufacturers have undoubtedly contributed to the growing demand for this type of construction. Modular construction's financial and practical advantages become evident to all stakeholders when they witness the accuracy and craftsmanship that can be attained in a controlled factory setting and understand its benefits in fewer onsite times and neighbourhood disruptions. Modularisation gives architects and designers complete freedom and creativity and provides cost and time efficiency for contractors and developers seeking high-skill, high-quality, large-scale projects such as major infrastructure, healthcare, and residential schemes.

5.4.3 Consumer Preference

Consumer preferences have been one of the main factors driving the growth of modular construction. Modular construction ensures that the real estate market is provided with cost-effective and environmentally-friendly housing solutions (Abdul Nabi & El-adaway, 2020). According to the report titled "Modular Construction

Market Size, Share, and Global Trend by Type, Material, End-User, and Geography Forecast till 2026", the global market for modular construction will grow steadily over the coming years, thanks to an increasing demand for speed in construction projects and a greater awareness of the effects of climate change. In recent years, the market has seen growth in using different types of modular construction, including relocatable and permanent modular construction, turning to e-commercial construction and building hotels, hospitals and others where one structure does it all. Modular construction provides a fast and effective alternative to the traditional methodology of on-site construction (Abdul Nabi & El-Adaway, 2021). Modular construction refers to using modules or 3D units, each individually manufactured off site in a factory-controlled environment, to create superior quality, more significant and cost-effective buildings in much faster timescales. It includes a variety of techniques, all of which can be referred to as modern methods of construction, from volumetric construction, in which 3D unity is stacked and adjoined, to panellised or flat pack construction, where panels and components are transported to the project site and assembled there.

On a broader scale, modular construction and other off-site construction methods fit in with the more comprehensive construction industry strategy of standardising and automating design and production, promoting greater efficiency within project delivery, closer community engagement and, ultimately, fostering a more sustainable industry. Various technological and procedural advancements and increased focus on better project planning underpin this approach. For example, implementing BIM, a sophisticated 3D model-based process that provides digital descriptions of a build's physical and functional characteristics, enables a more innovative, more efficient production management process from design to construction (Wuni & Shen, 2021).

Modular construction has several positive implications for a project thanks to its more sophisticated nature—from improved health and safety and reduced waste to increased quality and programme certainty. The application of the technology can potentially improve supply chain management across the whole construction industry, reducing single-supplier dependency and creating a fairer environment for competition (Choi et al., 2020). Despite modular construction offering the potential of a fully integrated and dynamic supply chain, the industry is still evolving from a gradualist mindset rooted in traditional methods. It is recognised that a step approach is necessary and indeed desired, allowing for the progressive elimination of waste and inefficiency. This, coupled with the need for large-scale primary investment, will see the full benefits of modern methods of construction realised in time.

5.4.4 Disaster Recovery

The recent disasters and emergencies, such as the COVID-19 pandemic, have highlighted the importance of modular construction in providing quick and effective emergency housing solutions. Conventional construction projects experienced significant disruptions due to the pandemic. Social distancing measures and lockdowns closed construction sites, and the global construction output fell by 4.1% in 2020 (Iqbal et al., 2021). In contrast, modular construction adapted more effectively to the new working conditions. Most of the construction work for modular projects is done in a factory, where it is easier to enforce social distancing measures and maintain a controlled working environment. As a result, the average modular project in the United States continuously grew from 156 housing units in 2019 to 210 units in 2020.

The potential applications of modular construction in disaster recovery are not limited to pandemic situations (Jiang et al., 2021); Ogunnusi et al., 2022). For example, in 2017, after Hurricane Irma, the Florida Keys needed 1300 new homes. A modular home manufacturer in the state was able to coordinate factory-based production to supply housing parts to the site, allowing for a much quicker rebuilding process than traditional construction would have allowed. In reality, many believe that the best example of how modular disaster recovery can be achieved originated from the United Kingdom. In 2013, after Typhoon Haiyan hit the Philippines and left many homeless, a student at Loughborough University won the Royal Institute of Chartered Surveyors' Pioneers in Innovation award for designing a rapidly deployable modular house. The innovative aspect of the design was the use of interlocking joint technologies, which allowed the house to be easily packed into a standard-sized shipping container. As a result, dozens of these homes could be loaded onto a single cargo ship and shipped to the disaster area very quickly. This example illustrates how modular disaster housing can change the landscape of international disaster relief efforts, and many hope that future initiatives will further expand the use of modular technologies in disaster recovery projects globally.

5.5 Economic Analysis of Modular Construction

Thai et al. (2020) opined that rather than being a disruptive force in the building industry, the adoption of modular construction represents the next evolutionary step, moving from an artisanal building to a modern methodology. Modernist architects and builders will recognise the potential of modular building techniques and a renewed avant-garde outlook. The close relationship between modular construction and computer-aided design is also evident. Using 3D models in the design process leads to significant efficiencies in building projects and gives clients a better understanding of proposed developments. Modular construction aims to provide better client value and do it more quickly (Chen et al., 2021). The approach works well in achieving such objectives as reducing waste and using materials efficiently. It also evidences a progressive shift in focus away from focusing on individual buildings as isolated projects with all of the associated logistical and resourcing difficulties and towards seeing them as products that are developed as part of a set (Wuni & Shen, 2020).

A building's life cycle dovetails with that of other similar projects. Moreover, one of the essential things about modular construction is that it helps to overcome the skilled labour shortage facing the construction industry (Gao et al., 2022). After the recession, there has been a significant disparity between the supply of skilled workers and the demand for their services in the market. This led to increased wages, which, in turn, hurt many projects. With the efficiencies and time savings that modular construction can bring, there is less reliance on labour. Therefore, more projects can be completed within the market rate for labour costs (Hossain et al., 2020). This factor could potentially significantly impact opening up new opportunities in places where, at present, building projects might be considered economically unviable owing to the high cost of labour. The potential economic benefits of modular construction are vast and varied.

The purpose of this sub-section is to serve as an overview of the economic aspects of modular construction, including both the challenges and potential rewards of integrating this method into the construction industry.

5.5.1 Cost Comparison with Traditional Construction

Owing to the increase in material costs and the pressure of green initiatives, many construction companies, developers, and architects are looking for green construction approaches. This will curb global demand for resources and research and use alternative methods that are both better for the environment and more cost effective (Pervez et al., 2022). This aspect provides an excellent opportunity for modular construction; however, careful research is needed to verify the extent of the advantages claims. Moreover, since most of the modular units are designed to be utilised as places such as wards in hospitals or study bedrooms in student accommodation, the design is fundamentally repetitive and much of the requirements could be satisfied by providing standard designs (Salama et al., 2021). Such designs take advantage of the volume of materials purchased and reduce the material waste from production. Veiskarami (2020) pointed out that around 19% to 30% of material waste in traditional construction could be reduced in modular construction. Therefore, the cost savings are so evident that materials such as steel, lumber, cement, and the like could save at least 10% of steel and 20% of lumber, which could be used for future projects or be consumed in national resources.

Zhang et al. (2020) analysis explains that the two main reasons for cost saving in the modular construction process over traditional construction in terms of expenses are economies of scale and reduced material usage. Modular construction companies would realise economies of scale in the long run. Since the companies are trying to maximise profit by simultaneously building several similar types of modules in the factory, the cost of creating the jigs and the equipment utilised to build the manufacturing modules could be minimised. Therefore, the unit cost of modular production decreases as the degree of standardisation increases and the number of similar types of work done. Through this long-term cost saving, the company can offer the client a cheaper price. This leads to a reduction in the whole construction cost by owners. The cost comparison analysis explains that the average cost saving per square foot of modular construction over traditional construction is about 13%. This is confirmed by averaging the total savings of $11,714,000 over the seven projects and dividing it by the average size of 62,705 square feet of the projects. Traditional construction has been proven to require many human and financial resources, including cost, labour, and time. Design and construction periods, costs, and other stages of investment in construction impact the national economy and the interests of investors. Modular construction is expected to offer cost, schedule, and aesthetic advantages over traditional construction.

5.5.2 *Labour Cost*

Another major factor in deciding whether to use modular construction is labour cost. Modular construction relies heavily on assembly line production in a factory as opposed to the traditional method, which requires continuous and simultaneous multiple trades to work on the site. Although labour cost varies significantly across different states in the US, it is widely recognised that construction labour is more expensive in metropolitan areas. A comparison of the average wages of construction labour between states from 2012 to 2016 reveals that in states such as Massachusetts, New York, and Pennsylvania, the average wages of construction labour are significantly higher than in North Carolina, Tennessee, and Texas. Also, according to the Bureau of Labor Statistics, New York construction wages experienced an increase of 10% from 2015 to 2016. This is a significant concern for traditional methods of construction, where construction labourers need to work indoors and often above ground in metropolitan areas. In the traditional methods, there are substantial requirements for more experienced labourers in order to deliver quality and safety. As a result, skilled labourers using traditional methods can demand a premium for their services.

In contrast, one main advantage of modular construction is that it can take advantage of lower-cost labour in rural outskirt areas while still maintaining high levels of efficiency. Also, the labour engaging in modular construction can be less skilled owing to the work's repetitive nature, which can be used to mitigate the lack of skilled labour in the industry. The repetitive nature of the modular construction is also attractive for potential business ventures. As the construction method is modular, it relies mainly on the efficiency of the factory assembly line by minimising the idle time of each labour and material. It is considered a more technical and capital-intensive investment than a labour-specific approach. On the other hand, the traditional method involves more of an on-site labour management approach, which is often more flexible and relies heavily on the decisions and coordination made by the site managers and trades. Taking advantage of the lean manufacturing concept, just-in-time production and delivery, widely used in aerospace and automotive manufacturing, can also be applied to modular construction. These lean tools provide systematic methods for waste elimination; the potential usage of these tools in modular construction draws attention from various professional bodies, and it is deemed that modular construction can promise greater flexibility in meeting customers' demands.

5.5.3 Schedule Acceleration

Schedule acceleration refers to the idea that a modular construction project can be completed in a shorter period than a traditional construction project. There is a much smaller lead time between the start of a project and when it is fully initiated in terms of construction and ordering materials. Because units can be manufactured in a factory while site work is being done, much of the on-site and off-site work can be done simultaneously, and the project duration can be shortened. The primary reason for schedule acceleration in modular construction is the parallel work between the building and site work (Wuni & Shen, 2020).

When talking about traditional construction, the lead time is much longer because if the same example of building a hospital is taken, for the foundation, we have to wait until site work is done. Then, when the foundation is done, we can build the main structure of the building and so on. This is why, in the built environment, civil and infrastructure engineers always want to devise innovative methods to shorten the lead time so that projects can be finished faster. The factory condition and mass customisation mentioned above will also help increase the schedule acceleration level, thus allowing the project to be completed even faster. Abdul Nabi & El-adaway (2020) argued that most of the pressure faced for project completion within a given time frame might lead many to change and try new construction methods, such as modular construction in the construction industry, to save time and complete a construction project more efficiently.

5.5.4 Waste Reduction

In traditional construction, excess materials are often left around the site for weeks and sometimes months. These materials are collected and taken to a landfill after the project completion. With a factory-based approach, unused materials can be more easily controlled and recycled efficiently. If materials are wasted, they are sent back to the factory for recycling and not thrown away on top of a building site. Estimates suggest that factory-based working produces around 90% less waste. This also means there is much less risk of damaging the surrounding environment and ecosystems on or around a building site. Coelho & De Brito (2012) opined that the waste from traditional construction also has a significant environmental impact. Building waste comprises two primary components: many heavy materials and a mixture of thousands of toxic chemical products such as adhesives, coatings, and solvents. When these waste products are sent to a landfill, it has been identified by the Environmental Protection Agency that there is a risk that these hazardous chemicals contribute to pollution in the air, soil, and groundwater in surrounding areas. Also, Herrmann et al. (2020) revealed that there are risks that some of the chemicals may reach drinking water supplies. If construction companies endorse a factory-based method, it could be seen as effective in reaching targets to reduce landfill by up to 90%

and reduce health and safety risks to workers and the public (Ibrahim et al., 2022). Moreover, a modular and sustainable approach to construction supports the aims of the Waste and Resources Action Programme (WRAP), a government organisation aiming to promote sustainable resource use and minimise the impact of waste on the environment. Economic incentives for modular construction include the reduction of wastage.

Owing to the accuracy in construction technology and the automation of processes, such as the quantity of materials ordered and the mounting of fastenings, there are fewer margins for errors (Greer & Horvath, 2023). As a result, excavation and groundworks for multi-functional foundations create little waste, manufacturers have more precision to optimise and reduce the use of materials, and automated fastening processes reduce the risk of loose components and excess 'spare parts'. Modern methods that favour minimised waste and sustainability mean that modular constructions should have advantages. First, it reduces the monetary costs for clearing away and disposing of the mass materials left over after construction, and it can also help businesses meet strict environmental targets (Ishirugi, 2023). This could be by simply removing landfills as a possibility for potential sites or by cutting back on damaging pollution to the local ecosystem by reducing waste. Second, projects can be accelerated by adopting offsite modular construction. By minimising the chance of shortages of materials and knowing exactly how much is required for a construction project, more precision planning for different stages of a build is possible. Moreover, site-based activity can be run in parallel, minimising overall production time. Third and finally, a modern construction approach could present new business opportunities. If the industry continues to shun traditional construction and look for newer, more environmentally friendly solutions when it comes to commercial developments, a 'greener reputation' could be developed by companies, encouraging collaboration and project-sharing for particularly innovative or large-scale construction. Economically, not only will the development of modern methods expand profits and encourage long-term investment, but there may also be a push for future research into the improvement and versatility of modular construction.

5.5.5 *Life-Cycle Cost Analysis (LCCA)*

Whole-life value thinking involves taking a longer-term view of the value offered by different products or processes (Bennett & Mayouf, 2021). Essentially, it is an approach that treats a project or process not as a one-off activity but as an ongoing, continually developing system with inputs and outputs. The aim is to maximise the efficiency and value of these outputs while minimising the inputs, and hence the costs. This has the potential to make the most effective use of resources and increase productivity, which benefits the project and the wider industry. By aiming to provide a long-term forecast of the useful life of a built asset, whole-life value thinking within a Life Cycle Cost Analysis (LCCA) can guide decisions regarding design, material, and construction options. In modular construction, LCCA is often used to compare the likely total costs of off-site manufacturing with traditional building methods (Alaloul et al., 2022). This helps clients and construction teams decide which construction method to use based on long-term value for money. Heidari et al. (2024) revealed that carrying out an LCCA at the planning stage makes it possible to explore the potential cost benefits of off-site manufacture and make an early comparison with traditional methods. This could help to ensure that a project which might appear more expensive to set up using offsite manufacturing represents better long-term value for money. Such an approach could also contribute directly to the client's whole-life value and sustainability criteria for the project.

As outlined by Roh et al. (2018) in Table 5.1, the private costs for construction in LCC were broken down into initial investment, operation costs, destruction, and other costs. Initial investment costs were subdivided into planning, design, and direct construction costs, while operation costs were subdivided into operation energy, maintenance, and general management costs. However, one of the challenges of LCCA is to ensure that all the potential project costs are identified and included in the analysis (He & Chen, 2021). For example, it is relatively straightforward to identify future costs such as heating and maintenance, but less easy to predict factors such as potential use changes for the building, which might involve extra costs. However, excluding a possible cost item now could provide misleading results from the LCCA and lead to a higher cost in the future that has not been planned for. LCCA involves identifying all the costs associated with a project, from the initial planning and design to the construction and eventual decommissioning of the building. It also involves adjusting these costs to allow for their value at different points in time. This is because a cost incurred in, say, 50 years has much less impact on a project's overall lifetime cost than one incurred next year. Therefore, future costs are said to have a lower 'present worth' than immediate costs. These adjusted costs are then combined to find the total lifetime cost of the project. LCCA is used to estimate the total costs of a project over its expected lifetime. This is especially important for building projects as the initial construction and materials costs are only a tiny part of the total expenditure. Most of the costs will be related to maintaining and operating the building throughout its life.

5.5.6 Financing and Funding Models

The study of Razkenari et al. (2020) highlights how important it is to link financing models with more significant economic policy and regulatory frameworks since it suggests that a financial commitment from the government would boost private investment in modular building. However, it is widely agreed that the long-term success and increased uptake of modular construction practices depend on establishing suitable, large-scale funding mechanisms. In this context, the industry professionals interviewed expressed support for the idea that governments and policy-makers could play a key role in promoting modular construction by offering financial incentives, such as tax breaks for modular projects or the provision of national or

Table 5.1 Life cycle cost categories

	Classification		Items
Initial investment costs	Planning and design costs	Planning costs Design costs	Basic design costs Execution design costs
	Direct construction cost	Market costs Labour cost Miscellaneous expenses	Materials, equipment, others Labor costs included in construction costs Various indirect costs, supervision costs, etc
Operation costs	Operation energy cost	Electricity bills Gas bills District heating and cooling bills Water and sewage bills	Electricity bills for air conditioning/hot water supply Gas bills for air conditioning/hot water supply District heating and cooling bills Water and sewage bills
	Maintenance costs	Repair parts cost Replacement costs	Repair and consumable parts costs Replacement costs, such as equipment and piping
	General management cost	Taxes Insurance premiums Labour costs	Acquisition and registration taxes on buildings Fire insurance premium Labour cost for general management
Destruction cost	Destruction cost	Destruction costs Residual values	Destruction costs of a building Residual values of a building
Other costs	Support funds	Various support funds	Support funds for electricity provider

Source Roh et al. (2018)

regional funding programmes specific to off-site construction methods. A modular construction expert, Lee Holmes, commented, "If the government were to try and encourage more modular builds, then I think it would be successful. If there were a certain pot of funds that's for modular builders, then the banks would be more likely to lend, and the financial institutions would be more likely to back you because they can see there is money there, and they want to get their share of the market."

Recognising that modular construction is a novel and relatively new idea is crucial for lenders, investors, and building experts (Abdelmageed & Zayed, 2020). This creates uncertainty regarding the durability and resilience of modular construction and the expected return on investment. For example, investors may be deterred from funding modular construction projects if they believe the cost savings and other financial benefits are less significant than the industry claims. Similarly, lenders may be hesitant to provide loans for modular construction if there is a widespread perception that the resale value of modular buildings is lower than that of traditionally constructed buildings (Jones & Stead, 2020). Various financing and funding models are available for modular construction projects. Despite that, developers and construction companies have found securing large-scale, long-term funding for such projects challenging. This section explores the reasons for these difficulties and discusses various funding options suitable for modular construction projects.

5.5.7 Cost–Benefit Analysis

Cost can be defined as the amount of actual or notional expenditure incurred on or required for a given activity or to achieve an objective. Conversely, benefit refers to the amount of advantage received over expenditure. The aim is to maximise the differences between benefits and costs. In the context of the cost–benefit analysis of modular construction, the costs are any drawbacks or limitations, and the benefits are their advantages. Economists have developed numerous approaches to assess economic benefits and costs in different contexts. These range from relatively simple and informal cost–benefit analyses to highly sophisticated and technical evaluation methodologies. Ultimately, they aim to help decision-makers have as much information as possible about a project's potential costs and benefits. However, a proper cost–benefit analysis depends on comparing modular construction against traditional construction methods. Specifically, the cost comparison with traditional construction will highlight the potential cost savings of modular construction. This analysis will give industry professionals and clients an idea of how modular construction is more efficient and will save money for the clients. The cost–benefit analysis will also determine the benefits of modular construction over older construction methods. The critical point is that the cost–benefit analysis will provide several angles to evaluate the overall economic rationale of using modular construction over more traditional construction methods.

Furthermore, another main focus within the cost–benefit analysis of modular construction is labour costs (Wuni et al., 2020). For many projects, labour costs will be a critical factor in determining the overall economic viability of modular construction. This is because modular construction occurs mainly within factory settings; workers are roughly 60 to 80% more efficient than those performing similar work on a construction site. As a result, labour costs tend to be lower for modular construction projects. Also, reducing the time needed to complete a project facilitated by modular construction reduces overall expenses on construction labour (Pervez et al., 2022).

Schedule acceleration is another critical concept and benefit of modular construction explored in the cost–benefit analysis. As modules are constructed off-site concurrently with other on-site foundation work, projects can be completed faster using modular methods. This strategic advantage of saving time can lead to considerable cost reductions due to decreased finance costs, inflation, and other costs associated with prolonging a project over a more extended period.

5.6 Matching-Up Modularity Methodology Application in the Construction Industry

To stimulate modularity methodology within the built environment, contemporary modular construction has entered a new era of innovation owing to technological advancements transforming the industry (Abdelmageed & Zayed, 2020). Technologies such as 3D printing have revolutionised the construction process by enabling the precise layer-by-layer deposition of building materials. In modular construction, 3D printing can create complex and customised components with unparalleled precision (Gosselin et al., 2016). This technology allows for intricate architectural designs that were once difficult or expensive to achieve using traditional construction methods. Additionally, 3D printing, automation and robotics can reduce material waste and construction timelines, making it an environmentally-friendly and cost-effective solution for modular projects. To this end, this study aims to assess the research on modular methodology application in the built environment to draw knowledge from previous studies. In contrast, the current study focuses on future applications related to innovative materials, digital twins, land use maximisation, and circular economy. This is done by identifying the published articles on modularity methodology in the built environment from various countries and authors while considering the most cited articles alongside the sources of the articles.

5.6.1 Innovative Materials

These new sustainable materials and efficient technologies can significantly influence designs and construction in the built environment. Applying these materials to the physical environment opens up possibilities and may solve several environmental issues. This, in turn, will enhance knowledge transfer in the construction industry. As the study of Papakonstantinou et al. (2021) mentioned, sustainable development calls for new, innovative solutions that integrate environmental, societal, and economic considerations in the design of products, buildings, and infrastructure. To put these innovations into practice, there is a need for an integrated approach involving architects, engineers, and builders, the development of new regulation codes, and the introduction of life-cycle assessment tools that support design for low environmental impacts, such as this material and new technologies. Puteri et al. (2022) also claimed numerous professional development activities, such as industrial secondments, placements, study visits, and practitioner talks, to enhance students' and researchers' career development and employability. The study postulated that further collaborations between academics and the construction industry can provide the necessary funding and facilitate commercialisation.

These innovative materials offer alternatives to conventional building materials. An example is clay-based materials, called agricultural waste-derived low embodied energy materials. These are made from local agricultural waste and have different surface qualities corresponding to various waste used (Hernández-Rodríguez et al., (2022). The material is developed to have the highest recyclability rate, lowest embodied energy and carbon footprint, and highest material value. This material can be used in various ways, from interior design to facade design. Another innovative material is luminescent solar concentrators, which facilitate solar harvesting in the built environment. This material functions as a transparent overlay that absorbs sunlight and provides it to solar cells as a mixture of light at visible frequencies. It is used for pre-fabrication because it costs less to prepare solar panels for construction. Bio-concrete is another material that can repair its cracks (Adekunle et al., 2021). The innovation spreads spores and nutrients within the concrete; when water enters through the cracks, these spores germinate and produce limestone to heal the cracks.

5.6.2 Digital Twin

In the course of a bibliometric analysis of modularity methodology in the built environment, a notable linkage emerges between the application and the concept of digital twins, a significant trend in 2022. The exploration of modularity methodology within the built environment often intersects with the application of digital twin technologies, as Yevu et al. (2023) observed. This reflects a growing recognition of the synergies between modular construction approaches and advanced digital modelling. The bibliometric analysis revealed that scholars are increasingly investigating how digital twins can enhance and complement modularity in the built environment. Digital twins, virtual replicas of physical structures or systems, provide a dynamic and real-time representation of modular components and their interactions (Jiang et al., 2021). This integration allows for a more comprehensive understanding of the modularity methodology, facilitating improved design, construction, and maintenance processes. In addition, studies by Jiang et al. (2022) and Lim et al. (2020) explored how digital twins can optimise the planning and implementation of modular structures. This involves utilising digital representations to simulate modular components' assembly, performance, and potential modifications, fostering a more efficient and accurate construction process.

5.6.3 Land-Use Maximisation

In the current urban setting, the literature review must emphasise the importance and capability of land usage. Examining modularity methodology within the built environment aligns with the broader goal of efficiently utilising available land resources. This reflects an awareness of the spatial constraints and the need for sustainable urban development. Studies need to explore the contribution of modularity methodology to land use optimisation as there are currently scanty publications. Modular methodologies will offer advantages in terms of spatial efficiency by enabling the construction of prefabricated components that can be assembled on site, minimising the footprint required for traditional construction. This intersection underscores the relevance of modularity in addressing the challenges of urbanisation, where maximising land use is essential for accommodating growing populations within limited urban areas. It is suggested that current methods diversify only in terms of data inputs and constraints set; for example, parametric design would take a set of configurations, inputs and constraints and derive the optimal parameters in generating a modular solution. This is also in line with establishing the method to the essence utilisation of land for the built environment. This means that different methods and practices of this tactic attempt to maximise land use. However, the thesis aims to introduce an innovative and compromising approach to land-use maximisation using a case in practice.

5.6.4 Circular Economy

The world is currently far from circular, with an estimated 91% of resources recycled or incinerated after their first use (Joensuu et al., 2020). The built environment is a massive contributor to this linear nature of our economy, with a staggering 60% of the world's materials being used to construct buildings and infrastructure and significant elements of global energy use. This poses an incredible challenge to sustainability and the strain of an ever-growing population. Recognising the issue of intensifying limited resources, Ahmend et al. (2022) have proposed a novel framework to change from a linear to a circular economic model, as seen in Fig. 5.4. This is an alternative to a traditional linear economy (make, use, dispose) in which resources are kept in use for as long as possible, the maximum value obtained from them whilst in use, and then recovered and regenerated products and materials are at the end of each service life. This contrasts with a linear model involving a 'take, make, dispose' approach to extracted resources, materials made into products, and products used until they are disposed of. The cyclical nature of a regenerative model is inspired by the way products are reused in both natural systems (reabsorption) and industrial systems (such as car battery recycling). As a result, a circular economy minimises waste and avoids pollution through long-lasting design, maintenance, repair, reuse, re-manufacture, refurbishing and recycling. According to the Wrap, the UK's leading sustainability charity, moving to a circular economy could add an estimated £29

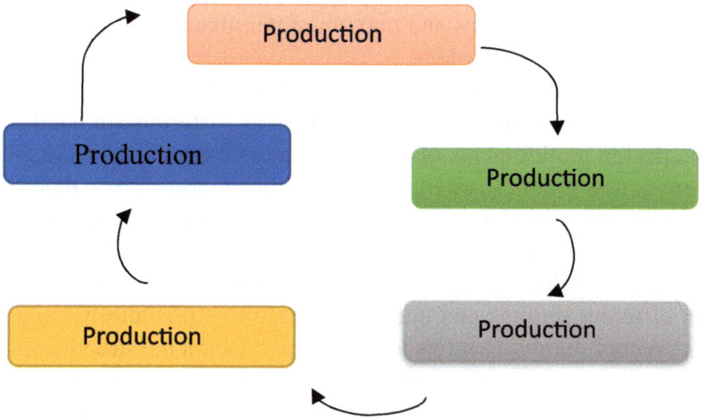

Fig. 5.4 Circular economy model. *Source* Ahmed et al. (2022)

billion to the UK economy and reduce greenhouse gas emissions by 50%. This has the potential to provide enormous economic, environmental and social benefits for the built environment, such as cost savings, reduced resource consumption and associated impact, and increased resilience to the risks of climate change and resource scarcity.

5.6.5 Robotics and Automation

Next-generation robotics technologies can significantly enhance all stages of the life cycle of the built environment. The increasing need for optimisation of the construction and assembly processes, the burgeoning field of semi-autonomous and fully autonomous construction robots, and the vision of self-reconfiguring ensembles of robotic agents with capabilities to adapt to changing environments are defining a set of research questions and design strategies for the utilisation of robotics systems in the construction and life cycle of buildings. Big data and advanced analytics can support robotics and construction automation by collecting and analysing information from diverse sources to optimise the operation of construction automation systems and to deliver new insights over the life cycle of the building. Additionally, Internet of Things (IoT) technologies are facilitating the installation of sensors in construction machines and intelligent stop solutions to further improve safety levels and machine reaction time in case of hazards. Automation can be direct when a robot receives control commands from a programmable logical controller or a computer numerically controlled system. However, plant control technology and automation systems are developing quickly so that the most different physical processes and units can be centrally or decentrally controlled flexibly, with high efficiency, cost-effectively, and with minimum risks, in the attempt to respond to ever-increasing needs for modularity, flexibility, and maximum possible automation in construction plants and robotic assembly systems.

In this regard, modularity, the possibility to create different combinations of control logics and kinematic rules to define the way materials are handled based on the specific design requirements of a component, is a common practice both in the development of new robotics systems and in the current research on modular robotics (Bi et al., 2020). Also, modern manufacturing engineering is based on the principle that a module's design and process planning should be developed and realised so that maximum flexibility and automatic adjustment can be reached for the scope of reuse in different settings and upon changes in manufacturing requirements. In this context, reconfigurability refers to the ability to adapt the structure and behavior of an automatic system's operational processes to changes in the environment. This adaptability also applies to the required operational tasks, based on different modular design methodologies. It is a crucial research area and remains an undeveloped field for construction automation, robotic systems, and the construction life cycle. These methods of construction automation and modern manufacturing engineering can offer the possibility to adapt the functionality of a robotic system and effectively reconfigure robots of a modular nature for different construction tasks during the life cycle of a building. This approach may also offer attractive solutions to cover the whole life cycle of the building more efficiently so that automation targets the construction stages and is planned to extend to maintenance and recycling activities. The capability to create work teams of autonomous yet adaptive robotic members, such as reconfigurable robots, actively supports the vision of need-oriented, self-organised automation in construction and assembly.

5.6.6 *Smart Buildings*

Innovative technology is now taking over the building industry, with intelligent buildings representing the primary way of applying digitisation to everyday life. The IoT is at the heart of it. The application of the IoT in buildings is connected to everything, such as the lighting, security, HVAC (heating, ventilation and air conditioning) systems, as well as the user's mobile device. Consequently, this creates a wide range of potential risks. If security measures are not appropriately implemented, unauthorised individuals can gain entry and disrupt the system or cause personal employee details to be leaked to the outside world.

According to Al Dakheel et al. (2020), three main characteristics define an intelligent building. Firstly, smart buildings should have an infrastructure that helps provide a comfortable, reliable and secure environment. Secondly, the building infrastructure should be intelligent, i.e., it should be able to react and respond to possible inputs, such as environmental changes. Finally, the building infrastructure should integrate many different technologies. These three main features thus aim to ensure that the physical building is efficient and cost effective and that its users get the maximum benefit from the technologies put in place. Intelligent buildings can offer extensive advantages, such as helping organisations use their resources more effectively, providing better and safer working environments and helping to minimise environmental impacts. These are good reasons why the global market for intelligent building technology is expected to grow to approximately $7 billion by 2024 and will continue to show significant growth year on year. Also, the application of modularity methodology in facilitating architectural-engineering systems and building products in line with the intelligent environment trend has been introduced and continuously developed recently. This particularly relates to the promotion of generative design, innovation framework, advanced digital technologies and process simplification on decision making for the design and construction of innovative building projects.

The segmented construction and prefabricated and modular design strategy is a crucial enabler for the engineers to tackle the new challenges generated by the multi-aspect and high-level complex bright environment. This methodology envisages and proposes a new way for industry professionals to make a revolutionary change in integrating various architectural-engineering parameters and building elements with modern IoT-connected innovative building principles. Therefore, from a software point of view, this concept has introduced new opportunities for different forms of software development. It is believed that adopting this methodology will improve the collaboration and knowledge sharing among various professionals such as architects, engineers, facility managers and end-users (Demlehner et al., 2021). However, this methodology's overall goal and concept still advocates integration internally and externally in the design and build process, improving the flexibility, scalability, modularity and interoperability of the building systems and minimising the dependency of a specific stage of the building life cycle.

5.6.7 Internet of Things

The Internet of Things (IoT) refers to the interconnectivity among devices over the Internet. This creates the possibility for devices to collect and exchange data with each other. In modular construction and the built environment, IoT can find its application in various ways. For example, intelligent and connected modular building elements can be equipped with sensors and actuators. Sensors can be used to monitor the state of the building or its environment, while actuators can be used to control elements of the building according to the data gathered by the sensors. These innovative moderators can autonomously establish a local area network (LAN) once they are physically connected. Then, a central master control unit can connect the modular unit to the external network. Also, there is much transparency in the operation and condition of the building, particularly for maintenance. Any deterioration in the building or any malfunction of the structure can be instantly notified to the responsible parties. This will significantly improve the life-cycle of the construction and building safety. Such bright modular elements can achieve self-adaption and optimisation by employing sharing and interchanging information with each other through IoT. The actual operation or the occupants' habits can be learnt and analysed. The building can adjust to different conditions and environments, making it more user-friendly. In summary, IoT provides the means for a modular building to interact intelligently with the environment. In this digital age, with the fast development of technology, utilising intelligent modular building envisaged by modularity methodology will soon become the traditional construction practice.

5.6.8 Nature-Based Solutions

Tiwary and Brown (2024) opined that applying modularity methodology in the built environment with nature-based solutions involves strategically integrating modular construction techniques with principles inspired by nature to create sustainable, resilient, and harmonious urban landscapes. Modularity methodology emphasises off-site fabrication and efficient assembly processes to streamline construction, reduce waste, and optimise resource utilisation. Nature-based solutions, such as biophilic design, green infrastructure, and sustainable urban drainage systems (SUDS), heighten modular construction by amplifying the positive environmental impact while enhancing the well-being of occupants and the broader ecosystem.

The significant aspect of this synergy lies in incorporating biophilic design principles into modular construction projects. Biophilic design seeks to reconnect inhabitants with nature by integrating natural elements, patterns, and materials into the built environment. Modular construction allows for the prefabrication of components that facilitate the incorporation of biophilic elements, such as living walls, green roofs, and natural light optimisation features. These elements enhance aesthetics and improve indoor air quality, reduce stress, and boost productivity, fostering healthier and more enjoyable living and working spaces. These nature-based solutions promote the development of green infrastructure networks within urban areas, which can be seamlessly integrated into modular construction projects (Ahamad et al., 2022). Green roofs, permeable pavements, rain gardens, and urban forests are examples of green infrastructure elements that can be prefabricated and easily installed using modular techniques. These features help mitigate urban heat island effects, reduce stormwater runoff, enhance biodiversity, and improve air and water quality, contributing to urban environments' overall resilience and sustainability.

Modularity methodology in the built environment will enable the rapid deployment of nature-based solutions in response to emerging environmental challenges. For instance, prefabricated modular wetlands or floating gardens can mitigate flooding, enhance water quality, and create a habitat for wildlife in urban areas prone to inundation or water pollution. In addressing pressing environmental issues, modular construction enhances the effectiveness and efficiency of nature-based interventions by providing flexible and adaptable solutions that can be quickly deployed and scaled up as needed. Nature-based solutions facilitate adaptive reuse and restoration projects, essential for conserving natural resources and preserving cultural heritage. The applicability of modular construction techniques allows for the repurposing of existing structures and materials, minimising demolition waste and reducing environmental impact. Nature-based solutions, such as urban green corridors or restored riparian habitats, can be integrated into adaptive reuse projects to enhance ecosystem services, biodiversity, and human well-being, transforming neglected or degraded areas into vibrant and resilient urban landscapes.

5.7 Summary

As more industries are pressured to move towards sustainable goals, increased globalisation denotes increased future building activity in less developed and disaster-affected regions. The lessons in global construction are often also lessons for the construction industry as a whole. With modular construction being a potentially cost-effective solution for both public and private sector clients, the essay has shown that there is much potential future demand, and it is hoped that research and method development do not repeat the same mistakes. A proactive approach and preparedness of modular methods are needed to cater to the increasingly complex and varying building requirements in various global markets. In this way, modular construction can shed its current stigma as a low-quality cost-cutting measure. By identifying these lessons, the thesis effectively demonstrates the significance of modular construction methods for future global construction projects and the industry. The findings are a stark revelation of the unprepared and ad hoc manner in which present global modular construction has occurred. In current economic climates and market trends, many international and national construction projects can be seen as a microcosm of global projects. They also seek cost-effective building solutions that do not compromise quality and functionality.

The review illustrates that present global modular construction often involves mere transposing of conventional onsite methods to seek time and cost savings. This is not the essence of offsite construction, and there is an error in the assumption that a method proven in the home market will also be suitable in a foreign market owing to the aforementioned complex requirements. Another lesson is that global construction projects present more demanding and complex requirements, such as differing national and regional building codes and standards, unfamiliar working environments, and availability of resources and skills. This thesis has shown that present modes of modular construction have not been able to cope with such complexity and have reverted to site-based improvisation. An example is the rework in design and dimensions of room partitioning in the case of the Taiwanese hospital project due to last-minute client requests and changes in building use. This has implications for the hoped increased planning and design activity offsite. Still, a change from onsite for module construction may involve future success in modularising repeating major building elements.

References

Abdelmageed, S., & Zayed, T. (2020). A study of literature in modular integrated construction-Critical review and future directions. *Journal of Cleaner Production, 277*, 124044. https://doi.org/10.1016/j.jclepro.2020.124044

Abdul Nabi, M., & El-adaway, I. H. (2020). Modular construction: Determining decision-making factors and future research needs. *Journal of Management in Engineering, 36*(6). https://doi.org/10.1061/(asce)me.1943-5479.0000859

Abdul Nabi, M., & El-adaway, I. H. (2021). Understanding disputes in modular construction projects: Key common causes and their associations. *Journal of Construction Engineering and Management, 148*(1). https://doi.org/10.1061/(asce)co.1943-7862.0002208

Adekunle, P., Aigbavboa, C., & Akinradewo, O. (2021). Intervention of biomimicry for sustainable construction: The use of bio-concrete. *Journal of Agronomy, Technology and Engineering Management, 2021*(5), 641–648. https://www.fimek.edu.rs/jatem

Ahamad, M. A., Arifin, K., Abas, A., Mahfudz, M., Cyio, M. B., Khairil, M., et al. (2022). Systematic literature review on variables impacting organization's zero accident vision in occupational safety and health perspectives. *Sustainability* (Switzerland), *14*(13). https://doi.org/10.3390/su14137523

Ahmed, Z., Mahmud, S., & Acet, D. H. (2022). Circular economy model for developing countries: evidence from Bangladesh. *Heliyon, 8*(5), e09530. https://doi.org/10.1016/j.heliyon.2022.e09530

Al Dakheel, J., del Pero, C., Aste, N., & Leonforte, F. (2020). Smart buildings features and key performance indicators: A review. *Sustainable Cities and Society, 61*. https://doi.org/10.1016/j.scs.2020.102328, 102328.

Alaloul, W. S., Alzubi, K. M., Malkawi, A. B., al Salaheen, M., & Musarat, M. A. (2022). Productivity monitoring in building construction projects: a systematic review. *Engineering, Construction and Architectural Management, 29*(7), 2760–2785. https://doi.org/10.1108/ECAM-03-2021-0211

Baduge, S. K., Thilakarathna, S., Perera, J. S., Arashpour, M., Sharafi, P., Teodosio, B., Shringi, A., & Mendis, P. (2022). Artificial intelligence and smart vision for building and construction 4.0: Machine and deep learning methods and applications. *Automation in Construction, 141*, 104440. https://doi.org/10.1016/j.autcon.2022.104440

Bennett, K., & Mayouf, M. (2021). Value management integration for whole life cycle: Post covid-19 strategy for the UK construction industry. *Sustainability* (Switzerland), *13*(16). https://doi.org/10.3390/su13169274

Bertram, N., Fuchs, S., Mischke, J., Palter, R., Strube, G., & Woetzel, J. (2019). Modular construction: From projects to products. *Capital Projects & Infrastructure*, 1–30. https://www.mckinsey.com/industries/capital-projects-and-infrastructure/our-insights/modular-construction-from-projects-to-products

Bi, Z. M., Miao, Z., Zhang, B., & Zhang, C. W. J. (2020). The state of the art of testing standards for integrated robotic systems. *Robotics and Computer-Integrated Manufacturing, 63*. https://doi.org/10.1016/j.rcim.2019.101893

Björnfõt, A., & Stehn, L. (2007). A lean perspective on off-site manufacturing in construction. *Construction Management and Economics, 25*(4), 415–427.

Blismas N., Wakefield R. (2009). Drivers, constraints and the future of offsite manufacture in Australia. *Construction innovation, 9*, 72–83. https://doi.org/10.1108/14714170910931552

Bender, D., Kim, H., & Woo, J. (2015). Framework for cost and performance benchmarking of modular construction. *Journal of Construction Engineering and Management, 141*(6), 04015007. https://doi.org/10.1061/(ASCE)CO.1943-7862.0000964

Chen, Z., Song, P., Wang, B., (2021). Carbon emissions trading scheme, energy efficiency and rebound effect—Evidence from China's provincial data. *Energy Policy, 157*, 112507.

Choi, J., Gu, B., Chin, S., & Lee, J. S. (2020). Machine learning predictive model based on national data for fatal accidents of construction workers. *Automation in Construction, 110*, 102974. https://doi.org/10.1016/j.autcon.2019.102974

Coelho, A., & De Brito, J. (2012). Influence of construction and demolition waste management on the environmental impact of buildings. *Waste Management, 32*(3), 532–541. https://doi.org/10.1016/j.wasman.2011.11.011

Dajsuren, Y., Van Der Putten, P., & Huisman, M. (2013). Modularity metrics for automotive software architectures. *Journal of Systems and Software, 86*(5), 1289–1305. https://doi.org/10.1016/j.jss.2012.12.025

Demlehner, Q., Schoemer, D., & Laumer, S. (2021). How can artificial intelligence enhance car manufacturing? A Delphi study-based identification and assessment of general use cases. *International Journal of Information Management, 58*, 102317. https://doi.org/10.1016/j.ijinfomgt.2021.102317

Dipasquale, L., Rovero, L., & Fratini, F. (2020). Ancient stone masonry constructions. In *Nonconventional and vernacular construction materials: Characterisation, properties and applications*. Elsevier Ltd. https://doi.org/10.1016/B978-0-08-102704-2.00015-9

Dubois, A., & Gadde, L. E. (2002). The construction industry as a loosely coupled system. *Construction Management and Economics, 20*(7), 621–631.

Egege, C. O. (2018). Off-site modular construction as a method of improving construction quality and safety. *International Journal of Structural and Civil Engineering Research, 7*(3), 259–268. https://doi.org/10.18178/ijscer.7.3.259-268

Enshassi, M. S. A., Walbridge, S., West, J. S., & Haas, C. T. (2020). Dynamic and proactive risk-based methodology for managing excessive geometric variability issues in modular construction projects using Bayesian theory. *Journal of Construction Engineering and Management, 146*(2), 1–16. https://doi.org/10.1061/(asce)co.1943-7862.0001747

Featherston, C. R., Ho, J. Y., Brévignon-Dodin, L., & O'Sullivan, E. (2016). Mediating and catalysing innovation: A framework for anticipating the standardisation needs of emerging technologies. *Technovation, 48–49*, 25–40. https://doi.org/10.1016/j.technovation.2015.11.003

Feldmann, F. G., Birkel, H., & Hartmann, E. (2022). Exploring barriers towards modular construction–A developer perspective using fuzzy DEMATEL. *Journal of Cleaner Production, 367*, 133023. https://doi.org/10.1016/j.jclepro.2022.133023

Gann, D. M., Salter, A. J., & Whyte, J. K. (2000). Design quality indicator as a tool for thinking. *Building Research & Information, 28*(5–6), 318–333.

Gao, Q., Liu, B., Sun, J., Liu, C., & Xu, Y. (2022). Trade decomposition of CO_2 emissions of global construction industries. *Engineering, Construction and Architectural Management, 29*(1), 502–522. https://doi.org/10.1108/ECAM-09-2020-0703

Ghannad, P., Lee, Y.-C., & Choi, J. O. (2020). Feasibility and implications of the modular construction approach for rapid post-disaster recovery. *International Journal of Industrialized Construction, 1*(1), 64–75. https://doi.org/10.29173/ijic220

Gosselin, C., Duballet, R., Roux, P., Gaudillière, N., Dirrenberger, J., & Morel, P. (2016). Large-scale 3D printing of ultra-high performance concrete—A new processing route for architects and builders. *Materials and Design, 100*, 102–109. https://doi.org/10.1016/j.matdes.2016.03.097

Greer, F., & Horvath, A. (2023). Modular construction's capacity to reduce embodied carbon emissions in California's housing sector. *Building and Environment, 240*, 110432. https://doi.org/10.1016/j.buildenv.2023.110432

Guerra, B. C., & Leite, F. (2021). Circular economy in the construction industry: An overview of United States stakeholders' awareness, major challenges, and enablers. *Resources, Conservation and Recycling, 170*, 105617. https://doi.org/10.1016/j.resconrec.2021.105617

Haas, C. T., Tucker, R. L., & Ballard, G. (2000). Prefabrication and preassembly trends and effects on the construction workforce. *Journal of Construction Engineering and Management, 126*(2), 83–90.

He, Z., & Chen, H. (2021). Critical factors for practicing sustainable construction projects in environmentally fragile regions based on interpretive structural modeling and cross-impact matrix multiplication applied to classification: A case study in China. *Sustainable Cities and Society, 74*, 103238. https://doi.org/10.1016/j.scs.2021.103238

Heidari, A., Peyvastehgar, Y., & Amanzadegan, M. (2024). A systematic review of the BIM in construction: From smart building management to interoperability of BIM & AI. *Architectural Science Review, 67*(3), 237–254. https://doi.org/10.1080/00038628.2023.2243247

Hernández-Rodríguez, M. A., Correia, S. F. H., Ferreira, R. A. S., & Carlos, L. D. (2022). A perspective on sustainable luminescent solar concentrators. *Journal of Applied Physics, 131*(14). https://doi.org/10.1063/5.0084182

Hong, J., Shen, G. Q., Li, Z., Zhang, B., & Zhang, W. (2018). Barriers to promoting prefabricated construction in China: A cost–benefit analysis. *Journal of Cleaner Production, 172*, 649–660. https://doi.org/10.1016/j.jclepro.2017.10.171

Hossain, M. U., Ng, S. T., Antwi-Afari, P., & Amor, B. (2020). Circular economy and the construction industry: Existing trends, challenges and prospective framework for sustainable construction. *Renewable and Sustainable Energy Reviews, 130*, 109948. https://doi.org/10.1016/j.rser.2020.109948

Hou, L., & Qian, S. (2016). Design for modular construction using BIM and knowledge management. *Journal of Construction Engineering and Management, 142*(6), 04016015. https://doi.org/10.1061/(ASCE)CO.1943-7862.0001113

Hsu, P. Y., Angeloudis, P., & Aurisicchio, M. (2018). Optimal logistics planning for modular construction using two-stage stochastic programming. *Automation in Construction, 94*, 47–61. https://doi.org/10.1016/j.autcon.2018.05.029

Ibrahim, C. K. I., Manu, P., Belayutham, S., Mahamadu, A. M., & Antwi-Afari, M. F. (2022). Design for safety (DfS) practice in construction engineering and management research: A review of current trends and future directions. *Journal of Building Engineering, 52*, 104352. https://doi.org/10.1016/j.jobe.2022.104352

Innella, F., Arashpour, M., & Bai, Y. (2019). Lean methodologies and techniques for modular construction: Chronological and critical review. *Journal of Construction Engineering and Management, 145*(12). https://doi.org/10.1061/(asce)co.1943-7862.0001712

Iqbal, M., Ahmad, N., Waqas, M., & Abrar, M. (2021). COVID-19 pandemic and construction industry: Impacts, emerging construction safety practices, and proposed crisis management framework. *Brazilian Journal of Operations and Production Management, 18*(2), 1–17. https://doi.org/10.14488/BJOPM.2021.034

Ishirugi, P. G. T. (2023). *Analysis of Modular Construction Techniques Advantages and Disadvantages*. Master's thesis, Universidade do Porto, Portugal.

Jaskolka, M., Pantelic, V., Wassyng, A., & Lawford, M. (2020). A comparison of componentization constructs for supporting modularity in Simulink. *SAE International Journal of Advances and Current Practices in Mobility, 2*(3), 1423–1432. https://doi.org/10.4271/2020-01-1290

Jeong, J., Jeong, J., Lee, J., Kim, D., & Son, J. W. (2022). Learning-driven construction productivity prediction for prefabricated external insulation wall system. *Automation in Construction, 141*, 104441. https://doi.org/10.1016/j.autcon.2022.104441

Jiang, Y., Li, M., Guo, D., Wu, W., Zhong, R. Y., & Huang, G. Q. (2022). Digital twin-enabled smart modular integrated construction system for on-site assembly. *Computers in Industry, 136*, 103594. https://doi.org/10.1016/j.compind.2021.103594

Jiang, Y., Yin, S., Li, K., Luo, H., & Kaynak, O. (2021). Industrial applications of digital twins. *Philosophical Transactions of the Royal Society A: Mathematical, Physical and Engineering Sciences, 379*(2207). https://doi.org/10.1098/rsta.2020.0360

Joensuu, T., Edelman, H., & Saari, A. (2020). Circular economy practices in the built environment. *Journal of Cleaner Production, 276*, 124215. https://doi.org/10.1016/j.jclepro.2020.124215

Jones, A., & Stead, L. (2020). Can people on low incomes access affordable housing loans in urban Africa and Asia? Examples of innovative housing finance models from Reall's global network. *Environment and Urbanization, 32*(1), 155–174. https://doi.org/10.1177/0956247819899557

Kamali, M., Hewage, K., & Sadiq, R. (2019). Conventional versus modular construction methods: A comparative cradle-to-gate life cycle assessment. *Energy and Buildings, 204*, 109479. https://doi.org/10.1016/j.enbuild.2019.109479

Khalfan, M. M. A., & Maqsood, T. (2014). Current state of off-site manufacturing in Australian construction. *Construction Innovation, 14*(4), 411–425.

Khoder, H., & Tommelein, I. D. (2017). Benefits and limitations of modular construction in improving project performance. In *Proceedings of the Lean and Computing in Construction Congress* (LC3) (pp. 275–284). https://doi.org/10.24928/JC3-2017/0126

Lessing, J., & Stehn, L. (2019). Industrialised construction: Definition and categorisation. *Construction Innovation, 19*(3), 427–446.

Love, P. E., Irani, Z., & Edwards, D. J. (2003). Learning to reduce rework in projects: Analysis of a firm's organisational learning and quality practices. *Project Management Journal, 34*(3), 13–25. https://doi.org/10.1177/875697280303400303

Lu, W., & Korman, T. (2010). Implementation of building information modeling using virtual prototyping technologies. *Automation in Construction, 19*(2), 158–172.

MacCormack, A., Rusnak, J., & Baldwin, C. Y. (2007). Exploring the duality between product and organizational architectures: A test of the "mirroring" hypothesis. *Research Policy, 36*(7), 993–1010. https://doi.org/10.1016/j.respol.2007.04.003

Moghadam, M., Alwisy, A., & Al-Hussein, M. (2012). Integrated BIM/lean base production line scheduling model for modular construction. In *Proceedings of the Construction Research Congress 2012: Construction Challenges in a Flat World*. ASCE Reston, VA.

Munaro, M. R., Tavares, S. F., & Bragança, L. (2020). Towards circular and more sustainable buildings: A systematic literature review on the circular economy in the built environment. *Journal of Cleaner Production, 260*. https://doi.org/10.1016/j.jclepro.2020.121134

Nahangi, M., Haas, C., & West, J. (2015). Automated real-time progress measurement using BIM-based spatial data. *Journal of Computing in Civil Engineering, 29*(4), 04014082.

Naim, M., & Barlow, J. (2010). An innovative supply chain strategy for customized housing. *Construction Management and Economics, 21*(6), 593–602. https://doi.org/10.1080/0144619032000134129

Noordzy, G., Whitfield, R., Saliot, G., & Ricaurte, E. (2021). Modular construction. *The Journal of Modern Project Management, 9*(2).

O'Connor, J. T., O'Brien, W. J., & Choi, J. O. (2015). Standardization strategy for modular industrial plants. *Journal of Construction Engineering and Management, 141*(9), 1–10. https://doi.org/10.1061/(asce)co.1943-7862.0001001

Olawumi, T. O., Chan, D. W. M., Ojo, S., & Yam, M. C. H. (2022). Automating the modular construction process: A review of digital technologies and future directions with blockchain technology. *Journal of Building Engineering, 46*, Article 103720. https://doi.org/10.1016/j.jobe.2021.103720

Pan, W., & Zhang, S. (2023). Digital technologies and industrialised construction. *Automation in Construction, 150*, 104840.

Pervez, H., Ali, Y., Pamucar, D., Garai-Fodor, M., & Csiszárik-Kocsir, Á. (2022). Evaluation of critical risk factors in the implementation of modular construction. *PLoS ONE, 17*, 1–27. https://doi.org/10.1371/journal.pone.0272448

Popescu, C. (2016). Soundness and completeness proofs by coinductive methods. *Journal of Applied Reasoning*, Published online.

Quale, J., Eckelman, M., Williams, K., Sloditskie, G., & Zimmerman, J. (2012). Construction matters: Comparing environmental impacts of building modular and conventional homes in the United States. *Journal of Industrial Ecology, 16*(2), 243–253. https://doi.org/10.1111/j.1530-9290.2011.00424.x

Rau, H. (2001). The concept of modularity in housing: Past, present, and future. *Journal of Housing and the Built Environment, 16*(1), 53–70.

Rausch, C., Sanchez, B., Haas, C., & Schreiber, F. (2020). Computational design approaches for modular construction. *Automation in Construction, 118*, 103274.

Razkenari, M., Fenner, A., Shojaei, A., Hakim, H., & Kibert, C. (2020). Perceptions of offsite construction in the United States. *Journal of Construction Engineering and Management, 146*(6), 04020044.

Salama, T., Salah, A., & Moselhi, O. (2021). Modular construction: State-of-the-art review. *Journal of Construction Engineering and Management, 147*(8), 03121005.

Schwinghammer, S., & Henseler, J. (2010). Modularity in product architecture: Review and outlook. *International Journal of Product Development, 11*(3/4), 276–299.

Shahtaheri, Y., Osanloo, M., & Karimi, B. (2017). Modular system design under uncertainty. *Resources Policy, 52*, 66–77.

Sharma, V., & Chandel, S. S. (2013). Performance and degradation analysis for long term reliability of solar photovoltaic systems: A review. *Renewable and Sustainable Energy Reviews, 27*, 753–767. https://doi.org/10.1016/j.rser.2013.07.046

Stephan, A., Crawford, R. H., & de Myttenaere, K. (2013). Towards a comprehensive life cycle energy analysis framework for residential buildings. *Energy and Buildings, 66*, 514–523. https://doi.org/10.1016/j.enbuild.2013.07.026

Song, J., Fagerlund, W. R., Haas, C. T., Tatum, C. B., & Vanegas, J. A. (2005). Considering prework on industrial projects. *Journal of Construction Engineering and Management, 131*(6), 723–733.

Vasawada, T., Singh, R., & Sawhney, A. (2015). Application of lean principles in modular construction. *Procedia Engineering, 123*, 61–69. https://doi.org/10.1016/j.proeng.2015.10.061

Wuni, I. Y., & Shen, G. Q. (2020). Critical success factors for modular integrated construction projects. *Journal of Cleaner Production, 262*, 121371.

Wuni, I. Y., & Shen, G. Q. (2021). Barriers to adoption of modular construction. *Engineering, Construction and Architectural Management, 28*(2), 460–484.

Zhang, S., Teizer, J., Lee, J. K., Eastman, C. M., & Venugopal, M. (2020). Building information modeling (BIM) and safety. *Automation in Construction, 111*, 103051.

Chapter 6
Conceptualising Modularity Methodology Model Adoption Framework

Abstract This chapter presents a conceptual framework for the adoption of modularity methodology in the construction industry, aiming to provide a structured approach for its effective integration into practice. The framework identifies the core components necessary for successful adoption, including technological readiness, regulatory alignment, stakeholder engagement, supply chain coordination, and cultural acceptance. These elements collectively form the foundation for a systematic and sustainable transition from traditional construction methods to modular approaches. The chapter outlines a systematic guide for implementation, offering step-by-step processes that address planning, design integration, manufacturing, logistics, on-site assembly, and quality assurance. It further highlights the role of continuous evaluation and feedback loops in ensuring adaptability and performance improvement over time. Strategies for advancing the modularity methodology in the construction sector are also examined. They consist of awareness campaigns, education, capacity building, training programmes and enhancing regulations. To foster an environment that is conducive to adoption, it is emphasised that government agencies, professional associations, technology companies, and construction companies must work together across sectors. Eventually, the suggested framework acts as a theoretical and practical instrument, addressing obstacles to broad adoption while directing stakeholders towards effective, economical, and sustainable modular construction methods.

6.1 Introduction

Jaganathan et al. (2013) carried out a study on modular coordination and industrialised building systems. This study focuses on modular coordination, establishing standard dimensions and rules for designing and constructing buildings. It explores how modular coordination improves construction efficiency, reduces costs, and facilitates prefabrication and modular construction. The study emphasises the importance of design modularity, component standardisation, and coordination in realising the benefits of industrialised building systems. Studies on modular design optimisation

P. Adekunle et al., *Modularity Methodology*,
https://doi.org/10.1007/978-3-032-12918-5_6

and application were explored by Kang et al. (2023) and Kharmandaa & Sayegh (2020). These studies investigate the optimisation of modular design in the context of structural systems. They explore methods for identifying and evaluating the modularity of different building components and systems. The studies highlight the benefits of modularity, such as improved constructability, reduced costs, and enhanced sustainability. They also discuss the challenges and considerations associated with implementing modular design. Another study, applying modular coordination principles in BIM by Banihashemi et al. (2018), explores integrating modular coordination principles with BIM. It investigates how BIM can facilitate the application of modularity methodology in the design and construction process. The study discusses the advantages of combining BIM and modularity, including improved design coordination, clash detection, and prefabrication. It provides practical recommendations for implementing modular coordination within BIM workflows.

Other empirical studies, such as those of Isaac et al. (2016) and Sharafi et al. (2017), examine the use of modularity in construction by analysing modular buildings and components. They investigate the factors influencing modularity in construction projects and assess the cost of modular systems' cost, time, and quality performance. These studies provide insights into the benefits and limitations of modularity, highlighting its potential for enhancing construction productivity and sustainability. The current study adopted the dependent variables per the findings of Bertram et al. (2019), which include supply chain, productivity, digitalisation, legal and regulatory considerations, sustainability and procurement methods.

6.2 Core Components of the Framework

Differing authors have postulated various theories and models depicting the acceptance of alternative building practices, such as modularity methodology. Wuni and Shen (2020) opined that the determinants of adopting modularity methodology are primarily influenced by the inherent features or characteristics of the discipline under review. With the development of various models and theories for innovative methods or acceptance, such as 'lean production theory', 'Queuing theory', 'Modularity theory', 'Control theory', 'Social cognitive theory', several studies have applied these different theories or models depending on the area or discipline being considered. The construction methodologies under review were relatively simple and also attributed to individual orientation as opposed to construction methodologies that are more sophisticated and complex; the measurement rationale was cross-sectional, and the majority of the studies were conducted involuntarily, enabling a problematic basis for making generalisations. These are some of the limitations that Bertram et al. (2019) observed after reviewing and testing the highlighted theories and models. Based on the backdrop of these identified drawbacks, Bertram et al. (2019) incorporated four constructs to determine the model for modularity methodology for the construction industry. These are supply chain, productivity, digitalisation, legal and regulatory considerations, sustainability and procurement methods. The current study supports and adopts these six constructs to determine modularity methodology for the construction industry.

6.2.1 Supply Chain

Storey et al. (2006) observed supply chain (SC) as a critical area of study and practice in modern business, encompassing the planning, execution, control, and monitoring of activities involved in moving goods and services from suppliers to customers. Various studies have contributed to understanding supply chain dynamics and the factors influencing its effectiveness (George et al., 2019; Hudnurkar et al., 2014). These studies focus on expectancy in the context of technology adoption. In the realm of supply chain, expectancy would relate to the belief that utilising SC systems or technologies will lead to improvements in supply chain performance. This also aligns with the idea that organisations adopt SC practices and technologies to enhance their supply chain efficiency, reduce costs, and improve overall performance. Sabri (2019) introduces the concept of supply chain fit, which refers to how well a system's ability can enhance the delivery of the supply chain. This notion is crucial in SC as organisations strive to align their supply chain processes, technologies, and strategies with specific operational requirements and objectives. A good fit ensures that the supply chain functions effectively and efficiently, meeting customer demands while minimising costs and risks. Hillman's et al. (2009) work on the resource dependence theory highlights the importance of relative advantage in adopting innovations. In the supply chain context, relative advantage refers to the perceived benefits of adopting new SC practices or technologies compared to existing methods. Organisations are more likely to embrace innovations in supply chain management if they offer significant advantages such as improved responsiveness, reduced lead times, enhanced visibility, and better coordination across the supply chain network.

6.2.2 Productivity

Classical economists such as Adam Smith, David Ricardo, and John Stuart Mill (Schabas, 2015) laid the foundation for understanding productivity in terms of labour and capital inputs. According to their theories, productivity increases when the same amount of inputs produces more outputs. This concept is often called the theory of diminishing marginal returns, where additional inputs yield decreasing marginal productivity. Furthermore, Frederick Taylor's principles of scientific management focused on optimising productivity through systematic analysis and standardisation of work processes (Giannantonio & Hurley-Hanson, 2011). Taylor proposed that productivity could be increased by scientifically determining the most efficient way to perform tasks and training workers accordingly. His time and motion studies aimed to eliminate wasted effort and improve efficiency. These studies emphasised the significance of employee morale and motivation in enhancing productivity. In addition, productivity is a construct in total quality management (TQM) theory that emphasises continuous improvement in all aspects of an organisation's operations, including productivity. By focusing on quality control, process improvement, and customer satisfaction, TQM aims to enhance productivity by reducing defects, waste, and rework. TQM principles include employee involvement, customer focus, productivity and data-driven decision-making.

6.2.3 *Digitalisation*

Diffusion of innovations theory (DOI), proposed by Rogers in 1962, explains how innovations spread through social systems. It identifies five adopter categories (innovators, early adopters, early majority, late majority, and laggards) and factors influencing adoption, including relative advantage, compatibility, complexity, trialability, and observability. DOI helps understand how digital innovations are adopted and diffused across different segments of society and organisations (Zhu et al., 2016). In contrast, the unified theory of acceptance and use of technology (UTAUT) integrates several information system (IS) theories, including the technology acceptance model (TAM), to explain technology adoption (Dwivedi et al., 2017). It considers factors such as perceived usefulness, perceived ease of use, external variables and internal characteristics. UTAUT provides insights into the factors influencing individuals' and organisations' adoption and usage of digital technologies. In addition, the technology-organisation-environment (TOE) framework, developed by Tornatzky and Fleischer in 1990, examines the factors influencing organisational innovation adoption (Baker, 2012). It considers technological factors, organisational factors, and external environmental factors. The TOE framework provides a holistic view of how organisations adopt and integrate digital technologies into their operations in response to internal and external factors.

6.2.4 *Legal and Regulatory Considerations*

An organisation's legal and regulatory compliance must ensure that the operational methodology they adopt complies with relevant laws and regulations, such as data protection laws, industry-specific regulations, and standards (e.g., ISO standards) (Pan & Jang, 2008). Studies such as those by Ioannou and Serafeim (2012) and Nayak and Jespersen (2022) highlight the importance of legal and regulatory frameworks about the system's ability to ensure better job performance. Legal and regulatory frameworks around policies and implementation are crucial in shaping users' perceptions and adoption behaviour. Behavioural works focus on the influence of legal and moral norms, risk perceptions, fairness, and legitimate processes on compliance motivations.

In contrast, regulatory cognitive works often examine how organisational actors understand and interpret regulatory requirements. Regulatory factors traditionally attract the attention of regulators and regulation scholars. This broad range of work typically examines the words used by regulators, how these words are transmitted, and how these words are filtered and distorted within organisational settings. Regulatory pressure is an umbrella term used to represent the enforcement of compliance requirements where a regulator enforces the over-arching rules of a regulatory system. Regulatory agents may apply reactive and proactive investigation tactics. Compliance requirements can allow companies to self-declare their compliance (voluntary compliance self-reporting) or require the regulator to certify that the company has complied. A key determinant of compliance choices will be interdependent with the regulatory context in which the requirement is set.

6.2.5 *Sustainability*

Zaharia and Zaharia (2021) opined that to understand what the "triple bottom line" (TBL) means for corporate management practice, it is crucial to know how companies use the concept. In the literature reviewed, the notion of the three dimensions of sustainability is often associated with TBL. If TBL represents a construct incorporating social, environmental, and economic elements of sustainable behaviour, then it is relatively straightforward for sustainability reports to indicate what has been done. Over the past decade, two distinct uses of TBL have been identified in the literature: operational values and TBL as a strategy (Olatunji & Olaoye, 2021). At a practical level, operational TBL reflects the notion of effectively considering three dimensions. If an organisation behaves in a way compatible with the idea of slavery, child labour, and lack of dignity in the community, no amount of social programmes will justify further operations. The extent of commercial impact on the social dimension of TBL receives media attention, and social media is assessed in light of this impact. However, more strategically, if the three dimensions are strategic values, management must maintain this balance specialised to long-term sustainability. At the level of a large corporation with a global footprint, this balance depends on the operations played out in different environmental and societal contexts (Székely & Vom Brocke, 2017). The TBL concept emphasises that the material interests of the organisation, people, and the planet can be aligned through simultaneous innovation in economic, environmental, and social strategies. Based on these considerations, organisations can formalise a roadmap related to sustainability. Elements associated with the social dimension cover not only organisational policies aimed at the well-being of workers and the communities of influence but also voluntary actions and cooperation agreements framed within models with a purpose of social responsibility, common cause, or supporting societal development. From the productive point of view, the implementation of activities linked to the social dimension should focus on the following aspects: the potential of the supply chain to reduce risks, manage costs, and generate social benefits as well as supporting community development and social sustainability projects.

6.2.6 *Procurement Method*

Employees' adoption of the procurement system is dependent on several factors, as observed by Cao & Wang (2022). These factors are taken from the studies on individual behaviour, known as social psychology, which mainly focuses on understanding the behaviour and intention of individuals and largely delimits the various theoretical models and theories. The popular theoretical models and theories are the Theory of Reasoned Action (TRA), the Theory of Planned Behaviour (TPB) and the Technology Acceptance Model (TAM2). Besides existing theories and models in the IT discipline, new models such as UTAUT have also been widely used. Procurement methods or procurement models are also related to organisational performance. Suppose an organisation can select the most suitable procurement model for their purchasing. In that case, it can be believed that the company's productivity, efficiency and financial situation will be affected. Therefore, the company's performance can be regarded as a dependent variable and the procurement methods (procurement models) as independent variables (Singh & Punia, 2011). The portfolio purchasing model, developed by Peter Kraljic (2019), classifies supplies into four quadrants based on their financial and risk impacts. It recommends different procurement strategies based on a supply item's classification and has been refined and verified in subsequent research. This model is a flagship in supply management and provides managerial insights for optimising procurement.

6.3 Systematic Guide for Implementation

There is a widely accepted view that research on adoption and the subsequent utilisation of construction methodology is regarded as one area in construction management literature that is most mature. However, there is still the inherent challenge of selecting an ideal or appropriate construct or model from the wide range of models when tasked with introducing innovative methodologies in construction organisations (Johnson et al., 2013). This is associated with the fact that different innovative methods apply to various disciplines. Hence, validating one model for a particular discipline might not necessarily fit into any other discipline. Martins et al. (2014) noted that there have been different applications of various theoretical models in diverse disciplines to provide an understanding and prediction on the validation of methodology adoption and usage determinants. Hence, many models and theories pose difficulties for researchers in selecting the ideal model for their study's strength.

Consequently, upon reviewing various models on the adoption of modular construction and innovative construction technologies and the subsequent examination of their strengths and limitations, this research is premised on the constructs of the project to product model (PPM) of Bertram et al. (2019) after being contextualised and modified. The strength and the flexibility of the theory project as a viable tool in acquiring the views and perceptions of individuals and organisations who have the responsibility of construction methodology. As a result, the study's conceptual framework examines the underlying strengths of the propelling drivers on individuals and organisations concerning modularity methodology for the construction industry. Table 6.1 indicates the constructs that comprise the PPM and match their emanating theories on acceptance of construction methodology.

There is a direct correlation between the objectives of the research and the constructs making the PPM. In contextualising the constructs of the theory, six constructs for modularity methodology for the construction industry were considered. These are categorically sectionalised into Supply Chain (SC), Productivity (PD), Digitalisation (DG), Legal and Regulatory Considerations (LR), Sustainability (ST) and Procurement Methods (PM). Based on the fundamental constructs outlined for modularity methodology from previous theories and models, the conceptualised model for this study aims to examine the relationship among supply chain, productivity, digitalisation, legal and regulatory considerations, sustainability and procurement methods, which are the resultants of the theoretical basis of this study, hence forming the exogenous variables; likewise their significance in modularity methodology for the construction industry which forms the endogenous variable.

The conceptualised model of modularity methodology for the construction industry is based on nine exogenous constructs (variables), which are Supply Chain (SC), Productivity (PD), Digi-talisation (DG), Legal and Regulatory Considerations (LR), Sustainability (ST), Procurement Methods (PM), Reassembly (RA), Maintenance (MT) and Cultural Perception (CP). The pictorial representation of the hypothesised model is depicted in Fig. 6.1. As the diagram indicates, the conceptualised model is a multidimensional framework comprising the earlier-mentioned exogenous

Table 6.1 PPM constructs and corresponding adoption model

PPM constructs	Matching construct	Matching adoption theory
Material flow	Supply chain	Resource dependence theory (Hillman et al.,2009)
Productivity	Productivity	Control theory
IT integration	Digitalisation	Technology acceptance model (Taherdoost, 2019)
Legal and regulatory Considerations	Legal and regulatory Considerations	Stakeholder theory
Sustainability	Sustainability	Lim's sustainability theory
Procurement strategy	Procurement methods	Resource dependence theory (Hillman et al., 2009)

construct (variables). The theoretical base for the a priori is the project-to-product model (PPM), as Bertram et al. (2019) postulated.

As stated earlier, the project-to-product model (PPM) outlines six exogenous constructs (variables): supply chain, productivity, legal and regulatory considerations, sustainability and procurement methods. However, owing to the area's unique nature of concern, it was necessary to include peculiar constructs that best project the intended construction methodology (modularity methodology), namely reassembly,

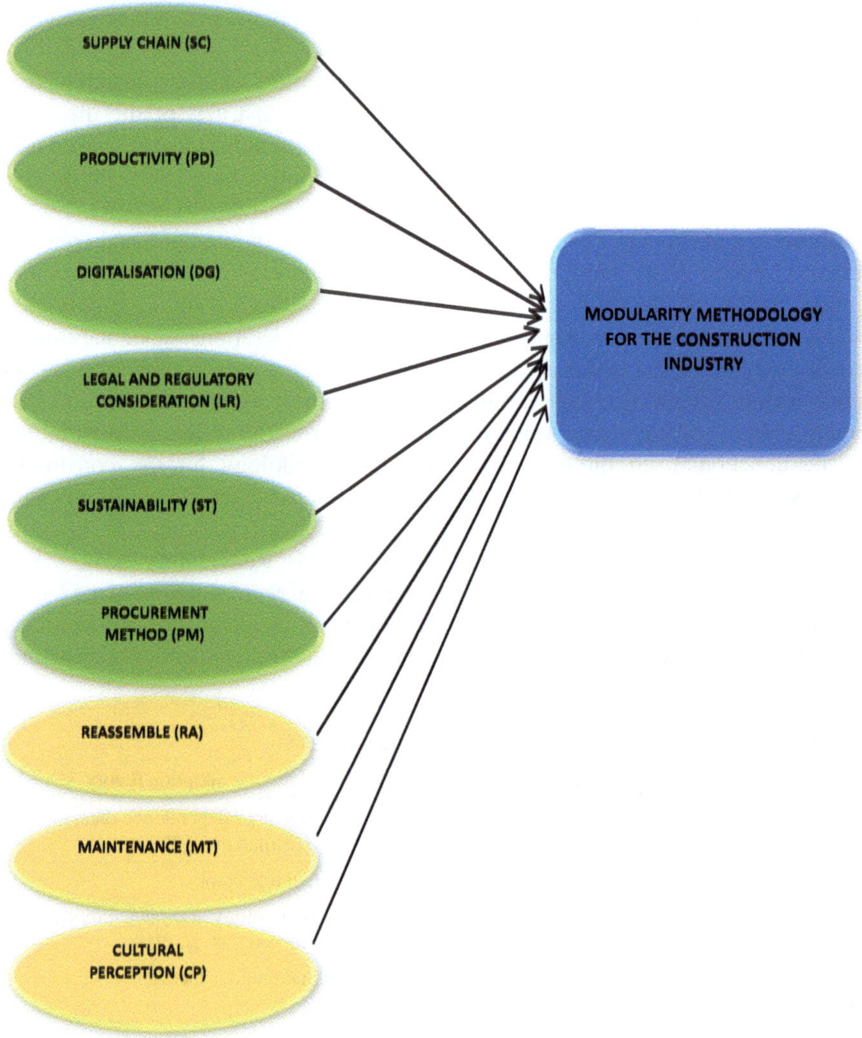

Fig. 6.1 Conceptualised modularity methodology model for the construction industry (Model 1.0)

maintenance, and cultural perception. Hence, the aggregate summary of all the exogenous variables is carefully crafted to suit its intended purpose: adopting modularity methodology for the construction industry.

6.4 Strategies for Promoting Modularity Methodology in the Construction Industry

Akinradewo et al. (2023) opined that capacity building and training programmes should be offered to construction workers, engineers, and architects to develop a critical mass of professionals competent in modular construction methods and design. These initiatives may involve the development of new courses or modules at universities, technical schools, and vocational training centres on modular construction methods and design or integrating these topics into existing curricula. Professional associations, construction companies, or manufacturers may also conduct training programmes for their members or employees. Developing a Nigerian standard or code for modular construction would also necessitate training programmes to familiarise industry professionals with the new requirements. Awareness and education campaigns should be designed primarily for public sector officials (at federal, state, and local levels), who are the key decision-makers in selecting construction methods. Elements of campaigns may also target the private sector, including firms and professional associations involved in the construction and real estate industries. The objective of the public sector is to inform key decision-makers about the range of available construction methods. This goal may be achieved through presentations or the distribution of information materials and may ultimately lead to demonstration projects using modular construction methods. For industry professionals, the objective is to build awareness of modular construction's benefits and convey the skills and knowledge necessary to work in the field.

6.4.1 Awareness and Education Campaigns

Awareness and education campaigns need to be organised to educate the public and construction professionals about modular construction and to communicate its advantages over traditional construction methods (Okoye et al., 2021). This would create a demand for a modern construction method (MMC) by highlighting its cost savings, time savings, and quality benefits. Primarily, campaigns will be aimed at the general public and construction clients who will need to be convinced of the efficacy of MMC, as they are the ultimate decision-makers (Salihu et al., 2021). Informational materials will be disseminated, and public seminars and exhibitions will be organised to educate these groups on MMC. Targeted marketing and information campaigns will also target specific government and private sector groups initiating construction projects to increase their awareness and understanding of MMC techniques (Aboginije et al., 2021). These campaigns will be carried out using brochures and videos outlining MMC's advantages, which can be accessed from government procurement websites. A public relations and media strategy will also be used to increase the profile of MMC in Nigeria and create a 'buzz' about the new techniques. This would involve getting prominent members of society to endorse MMC and using traditional and online media to get the message across to the population.

6.4.2 Capacity Building and Training Programs

The specific nature of training initiatives will evolve as the industry progresses through various development stages. Early on, training will be based on informing industry stakeholders about the benefits and methods of modular construction compared to traditional methods. Awareness campaigns and the development of educational materials can support this. Debugging myths surrounding modular construction will also be important. For example, many people within and outside the construction industry still associate modular buildings with low-quality relocatable classrooms. This leads to a stigma, which can be a barrier to acceptance of the method. Training programmes may also involve seconding workers to companies in countries with more developed modular construction industries (Luo et al., 2015). This will allow workers to develop skills in an environment where best practices are more established. The long-term vision is to create a training infrastructure that provides recognised qualifications in modular construction. High-quality training can attract people to the industry and increase labour productivity. Research projects which aim to develop an understanding of construction processes suitable for the Nigerian environment will also contribute to capacity building. An example was a research report by the Cement and Concrete Association of New Zealand, which investigated low-cost housing construction techniques using precast concrete floor and wall panels. This report identified that significant savings in both time and cost could be made compared to traditional construction methods. National Precast Concrete Association (NPCA) and Pre-Construction Information (PCI) also actively support research projects in the precast and concrete industry to improve understanding of technology and its application.

Packaging plants where these are manufactured are located in developed countries. An enabling environment that links local companies with worldwide markets will be a key to the success of capacity building. This can be supported through government matching of funds programs for research, training, or marketing initiatives. To set a benchmark in quality and technology, partnerships between international and local companies can be facilitated. This is seen as an effective method for developing technical skills and know-how. These partnerships can involve technology licensing arrangements—allowing overseas firms to build for the local market using their patented technology. This can be contrasted with joint ventures where two or more companies create a new entity to manufacture construction components. In the long-term development of the industry, skills must be localised. The skills development stage can also be supported by developing industry-specific curricula at local universities or training institutions often developed in conjunction with the industry. In the initial stages of industry development, workers with skills in more traditional construction methods will likely have to be retrained. Removing trade barriers preventing the entry of overseas construction professionals will also positively affect the quality of design and construction in the industry. The ultimate objective is to create a culture of continuous improvement in the industry.

6.4.3 Collaboration and Partnrships

Developing a widespread modular construction industry in Nigeria will require consistent collaboration and partnerships between numerous public and private stakeholders. Firstly, the ruling government must ensure that modular construction goals are integrated within all relevant government departments by forming an inter-ministerial taskforce to facilitate its promotion and adoption. This will provide a platform for various governmental organisations to align their policies and procurement decisions to support the modular construction industry. In the case of Nigeria for example, an MoU was signed between the National Universities Commission and REDAN to provide land for mass staff housing. It can be argued that the traditional construction method was selected as modular construction, which was not known or considered an alternative. A similar occurrence to the setting up of the International Housing Industrial Development Unit in the UK's seventies may have never been heard of as the knowledge and information was not there. At an industry level, there may be potential for cooperatives among companies where contractual work is given as a concerted effort to facilitate early direct experience for the national workforce. Finally, and possibly most importantly, there needs to be a consultative partnership with the communities where housing or infrastructure is targeted for construction. With modular construction in disaster relief housing, community-based trial projects and mock-up housing can allow affordable and versatile housing to take precedence over mass shelter, which may not have any permanency in providing housing solutions for those affected.

6.4.4 Awareness and Education Campaigns

Modularity methodology has long been viewed as a temporary building structure predominantly for the resource and construction industries. This perception results in many building projects defaulting to conventional site construction methods. The longevity of prefabricated buildings is misunderstood and not addressed in current code and planning regulations. A policy that recognises the difference between relocatable and permanent modular buildings and establishes approval and permit processes relative to the low-risk nature of prefabricated building techniques is required. Increase the use of prefabrication in the building sector by x% by a set date. Ensure all government-funded new buildings are built using prefabrication techniques. Encourage the reuse of building materials and the reduction of waste when building off-site. This is also important for countries importing prefabricated products be monitored by developing a sustainable building policy and rating system. A comprehensive review of the current code system and the policies impacting modular construction is required. Each geopolitical zone should have its own code system and policies. An integrated sustainable building policy and guidelines must be developed to consolidate these policies. The policy should promote high-performance modular construction and establish targets for increasing the use of prefabrication in both the residential and non-residential building sectors. Special consideration should be given to public procurement policy and the involvement of modular construction in rebuilding and repair work post-natural disaster.

6.4.5 Strengthening Financial Support Mechanism

The government often provides financial incentives for companies to innovate, usually through funding. This is often done without much consideration for the high level of fragmentation in the construction industry. As it has been identified that smaller builders and tradespersons are more inclined to innovate, funding initiatives must also target these groups. One way to do this could be to link payments with apprentice employment and training to ensure the money is injected into the industry. Any such initiatives must be carried out with thorough research and consideration of how the money can best be utilised, namely whether it is feasible for the change to occur, the expected improvement level, and the best way to measure the initiative's success. An informed and strategic approach is vital for successfully promoting innovation in construction.

Whilst the private sector is the critical driver of change in the construction industry, their initiatives are often too risky to consider under tight budgets and time constraints. More recently, public–private partnerships (PPPs) have emerged as a way for the government to share some of the risks and responsibilities of innovation in construction (Carbonara & Pellegrino, 2019). The Australian NSW Government, for example, has a Rapid Bridge Construction Program, which engages the private sector to build and test bridges using innovative methods. The most successful methods are then used in communities to replace bridges that need replacement. An example of a PPP, as seen here, would be the most effective way for the government to promote innovation in construction without taking on too much risk.

Easing insurance bonds for innovative projects would be another way for the government to reduce the risk of the private sector. Government ministers can lead and oversee the promotion of systematic innovation in construction to improve efficiency and productivity. They can often facilitate change within their departments, agencies, and industry. Adopting less traditional procurement methods can encourage innovation within public sector construction. Appointing key individuals such as 'innovation champions on projects or forming strategic alliances within the private sector will help drive innovation. Measures should also be implemented to evaluate the success of innovative programmes and the value they have added to the project to improve and learn from past mistakes continually.

6.4.6 Encouraging Research and Development

The undertakings of research and development (R&D) in the construction sector in both the public and private segments must be given incentives such as funding allocations, fiscal incentives, or grants, either directly from the government or through collaboration with local universities. Moreover, R&D institutions actively develop new materials, construction systems, and methodologies to improve productivity and cost-effectiveness and minimise the adverse impact on the environment. An example of a successful practice was the funding for the development of IBS conducted by CIDB from the year 1998 to 2008, which managed to increase the usage of IBS from 3.56% to 12.96% and helped to identify 73 building systems (CIDB, 2009). This also includes the need to develop a skill base for R&D activities in construction, which involves construction professionals, researchers, and policymakers relevant to the construction sector for efficient technology and knowledge transfer. There appears to be an apparent lack of R&D on modern construction methods (MMC), and most of the advancements are transferred from other sectors owing to the lack of a clear R&D and innovation strategy in the construction industry. For any country to possess a sustainable and efficient construction industry, it must encourage the development of new products, processes, and technologies. As the industry leader, the government must play a crucial role in formulating a conducive environment for R&D activities and promoting a culture of innovation.

6.5 Summary

This chapter focused on the conceptualised model for adopting modularity methodology for the construction industry, which is attributed to a multidimensional framework comprising nine constructs: supply chain, productivity, legal and regulatory considerations, sustainability procurement methods, reassembly, maintenance, and cultural perception. These variables are products of a well-reviewed literature review and the interview carried out for the research study. Furthermore, this section provides a comprehensive explanation of the theoretical foundation underlying the formulation of the conceptual model for adopting modularity methodology within the construction industry.

References

Aboginije, A., Aigbavboa, C., & Thwala, W. (2021). A holistic assessment of construction and demolition waste management in the Nigerian construction projects. *Sustainability (Switzerland), 13*(11), 1–14. https://doi.org/10.3390/su13116241

Baker, G. (2012). *Incentives and performance*. Harvard Business School Press.

Banihashemi, S., Tabadkani, A., & Hosseini, M. R. (2018). Integration of parametric design into modular coordination: A construction waste reduction workflow. *Automation in Construction, 88*, 1–12. https://doi.org/10.1016/j.autcon.2017.12.026

Bertram, N., Fuchs, S., Mischke, J., Palter, R., Strube, G., & Woetzel, J. (2019). *Modular construction: From projects to products*. McKinsey Global Institute.

Carbonara, N., & Pellegrino, R. (2019). Public–private partnerships for sustainable construction. *Sustainability, 11*(2), 1–17.

CIDB. (2009). *Construction industry strategy*. Construction Industry Development Board, South Africa.

Dwivedi, Y. K., Rana, N. P., Jeyaraj, A., Clement, M., & Williams, M. D. (2017). Re-examining technology adoption models. *Information Systems Frontiers, 19*(2), 1–20.

George, G., Haas, M. R., & Pentland, A. (2019). Big data and management. *Academy of Management Journal, 62*(2), 1–25.

Giannantonio, C. M., & Hurley-Hanson, A. E. (2011). *Managing human resources in project management*. Praeger.

Hillman, A. J., Withers, M. C., & Collins, B. J. (2009). Resource dependence theory. *Journal of Management, 35*(6), 1404–1427.

Hudnurkar, M., Jakhar, S., & Rathod, U. (2014). Factors affecting collaboration in supply chain. *Procedia—Social and Behavioral Sciences, 133*, 189–202.

Ioannou, I., & Serafeim, G. (2012). What drives corporate social performance? *Academy of Management Journal, 55*(4), 834–864.

Isaac, S., Bock, T., & Stoliar, Y. (2016). A methodology for the evaluation of building industrialization. *Automation in Construction, 65*, 44–52.

Jaganathan, S., Jeyaraj, K. L., & Chidambaram, S. (2013). Productivity improvement in modular construction. *International Journal of Engineering Research and Technology, 2*(4), 1045–1052.

Johnson, G., Whittington, R., & Scholes, K. (2013). *Exploring strategy* (10th ed.). Pearson.

Kraljic, P. (2019). Purchasing must become supply management. *Harvard Business Review Press.*

Luo, Y., Wang, S. L., Jayaraman, V., & Zheng, Q. (2015). Governing business process outsourcing. *Journal of Management Studies, 52*(8), 1210–1243.

Martins, L. L., Gilson, L. L., & Maynard, M. T. (2014). Virtual teams. *Journal of Management, 40*(3), 805–835.

Okoye, P. U., Ezeokoli, F. O., & Ugochukwu, S. C. (2021). Barriers to BIM adoption in Nigeria. *Engineering, Construction and Architectural Management, 28*(3), 1–19.

Olatunji, O. A., & Olaoye, A. J. (2021). Sustainable construction in developing economies. *Journal of Construction in Developing Countries, 26*(2), 1–19.

Pan, A., & Jang, W. Y. (2008). Determinants of supply chain management practices. *Journal of Manufacturing Technology Management, 19*(3), 291–308.

Sabri, E. H. (2019). *Lean and agile value chain management*. Springer.

Salihu, S. O., Jibril, J. D., & Waziri, B. S. (2021). Procurement challenges in Nigerian construction. *Built Environment Project and Asset Management, 11*(4), 1–15.

Schabas, W. A. (2015). *The international criminal court: A commentary on the Rome statute*. Oxford University Press.

Sharafi, M., Mortazavi, S. M., & Akrami, M. (2017). Productivity analysis of prefabricated construction. *Construction Economics and Building, 17*(4), 1–15.

Singh, S., & Punia, D. K. (2011). Benchmarking supply chain management practices. *Benchmarking: An International Journal, 18*(6), 1–20.

Székely, F., & Vom Brocke, J. (2017). What can we learn from corporate sustainability reporting? *Journal of Cleaner Production, 139*, 1117–1131.

Taherdoost, H. (2019). Importance of technology acceptance assessment. *Procedia Manufacturing, 32*, 1018–1025.

Wuni, I. Y., & Shen, G. Q. (2020). Critical success factors for modular integrated construction projects. *Journal of Cleaner Production, 262*, 121371. https://doi.org/10.1016/j.jclepro.2020.121371

Zhu, Q., Sarkis, J., & Lai, K. H. (2016). Corporate sustainability and green supply chain. *Journal of Environmental Management, 186*, 245–258.

Part III
Case Study

Chapter 7
Validation of the Conceptual Framework

Abstract This chapter focuses on the validation of the proposed conceptual framework for adopting modularity methodology in the construction industry. Validation was achieved through expert interviews designed to assess the framework's relevance, practicality, and applicability within diverse construction contexts. The selection of interviewees was based on their expertise, professional experience, and involvement in modular construction practices, ensuring that insights were drawn from knowledgeable and credible sources. The chapter also outlines the interview framework and approach, which combined structured and semi-structured questions to capture both standardised and in-depth perspectives. This method facilitated the collection of qualitative data that reflects expert opinions on the framework's strengths, limitations, and areas for refinement. Analysis of the data employed thematic coding, enabling the identification of patterns, emerging themes, and consensus points among experts. To ensure the credibility of findings, measures were adopted to enhance validity and reliability, including triangulation of responses, peer debriefing, and careful documentation of the research process. The validation process not only confirmed the robustness of the framework but also provided critical feedback for its enhancement, particularly regarding stakeholder collaboration, regulatory alignment, and cultural adaptation. The chapter demonstrates that the validated framework offers a sound, adaptable, and practical guide for advancing modularity methodology in the construction industry, strengthening its theoretical and practical contributions.

7.1 Introduction

The interview techniques involve using questions and conversations to elicit relevant knowledge from individuals or groups about the issues of concern (Murto et al., 2020). For many people, research is perceived as generating knowledge or information through a printed document. To these people, interviewing is just one more technique that should be utilised as a part of triangulation to use a mixed-methodological approach. The general purpose of interviewing is therefore to conduct surveys, use experiments, subject participant observers to the rigours of observation of a large

P. Adekunle et al., *Modularity Methodology*,
https://doi.org/10.1007/978-3-032-12918-5_7

number of participants, and aid in the recounting of personal events that had a measurable impact on the lives of members of a chance sample (Thelwall & Nevill, 2021). According to experts, a related misconception is that participants were often permitted to speak excessively, using jargon and specialised terminology that the interviewer neither understood nor had any realistic chance of comprehending. This can lead to communication barriers and affect the depth of insights gained during the interview process.

Hockey and Forsey (2012) explained that interviewing is one of the significant research strategies used by various researchers to collect data and historical information in the form of oral accounts. In some research practices, it is considered an ethnographic method rather than a method study in its own right. Rinaldo and Guhin (2022) believe the interview is part of participant observation and should be treated accordingly. It is part of the interaction among people in a social situation and is, to some extent, 'available' to the outside observer, mainly when the observer is a participant. In some traditions, particularly sociolinguistic traditions that make free use of conversational analysis, interview data has been used as one of the grounding trends for other theories. A peculiar set of transcription conventions has been developed to ensure that all the information gained from such data can be extracted, identified, and interpreted with the maximal suspension of the transcriber's assumptions about the categories involved. The interaction mechanisms in interviews can give researchers a more profound understanding, especially in helping them reach the tacit meanings of the informants about some issues and the meanings in everyday life (Yoo et al., 2022). Scheinin et al. (2021) argued that the data derived from interviews indicate not only the persons' conscious purposes and intentions in the plane of the questions asked but also their unconscious layers as they are asked to express what they expose in their thoughts and to 'narrate' themselves as if providing clues to the hidden or closed layers of the meaning. "With a judicious handling of its procedures," supported that only research conducted using open and in-depth interviews could be the "research that would allow the individuals to be heard" and that it could be the method that could convey the voices of the individuals (Hansen et al., 2022, p. 3). During the interviews, researchers become involved in the field they examine and reach informants' tacit understandings of the specific issues.

The advantages of using interviews involve the directness, the interviews' interactive communication mechanisms, the trustworthiness of the findings, and the employment of unique and individual features of the informants (Martin & Jamieson-Proctor, 2022). Interviewers' ability to deliver pre-constructed questions and instructions regarding participating in the inquiry, defining the contexts and determining the contents, and the possible length of the talk time in interviews contribute to the directiveness of the data collection process. Unlike participant observation, each party in interview-based studies finds an opportunity to understand the interviewer's purpose and objectives and reflect on what is happening during the interview interaction (Hosseini et al., 2022). Interviews are based on a bilateral and voluntary basis. The semi-structured interviews are expected to form structured research settings while guaranteeing freedom and trust to the parties for expressing their thoughts on specific matters to a great extent. However, as stated, a well-designed questionnaire already consists of structured interview questions. This suggests that the degree of structuring in interview-related data collection differs from the fully structured samples in analyses. In that respect, collecting natural and spontaneous data is a common goal in all research methodologies.

The interview is a field study research method which can be used as a structured tool or method for gathering data (Ruslin et al., 2022). The need to include interviews as a research methodology for this research project arose because of the flexibility in gathering data. The form the data collection took varied from project to project, and the topic of investigation, the timeframe, and the budget determined the choice of research tools. Still, it is essential to realise that, regardless of the chosen tools, they must be carefully selected and implemented. The research's ultimate objective and the researcher's skills are essential in decision-making. When properly selected and implemented, the quality of the research findings is assured (Islam & Aldaaihani, 2022). However, the results of the research are not always the only parameters that must be considered in the decision-making process. Time frame, money, and specific issues about the scope of the study are some of the other factors which push the decision-making.

In the social sciences, with the qualitative approach, the method used more frequently is the interview (Busetto et al., 2020). This technique provides a natural means of obtaining complete and authentic information about any phenomenon of concern. The interview, one of the qualitative research methods, is a technique that consists of a planned meeting in which the interviewers, using an interview guide, obtain a self-report from each of the respondents (Ruslin et al., 2022). In part, its design is based on the interactions that can occur in the interviewees' daily lives. These interviews usually occur in a single moment, and the interviewees act as the informants.

Many aspects need to be observed in the approach and the practice of qualitative research. Three types of interviews are part of the qualitative research approach: semi-structured, in-depth/unstructured, and structured. The purpose of the investigation and the research question are essential in preparing to research a social phenomenon. The first is the main reason for the study, while the second consists of what is to be known about the object of interest. After defining these aspects and determining the theoretical singularities about the object, one should consider the methodological research tools or the research techniques. Methods are the specific ways a study is developed, and their techniques are the procedures of concrete activities with which that research is applied. To best choose and apply these, the theoretical knowledge of those methods and techniques is fundamental, recognised by the denotative term of methodological procedures.

Structured Interview

The structured interview questions have multiple-choice responses; there is often a consistent format to encourage consistency in response coding, as employed by this study. The questions and responses are determined depending on the research question to collect relevant and measurable data (Røed et al., 2023). Using questions defined in advance, the researcher can avoid generating and asking questions during the interview. This way, it can be ensured that all the respondents are treated in the same manner, i.e., the formulation of a problem. Respondents in structured interviews can not only be passive answer givers but actively participate through involvement in the researcher's demand for a specific type of word or phrase as an answer to a predetermined question. In the field of communication, for example, Rankin (2023) stated that structured interviews can be applied not only in the research of organisational communication (as one part of human communication) but in the research of other segments of communication studies as well. In a structured interview, predetermined, standardised questions are used, and the order of asking these questions and the predetermined set of answers is defined in advance (Galana et al., 2023). Structured interviews are researcher led and are characterised by pre-fixed questions with pre-coded answers. In order to ensure reliability and to control the possible influence and guidance of the interviewer, the questions are predetermined and standardised, implying that the interview questions remain the same for every respondent. Although a pre-designed and fixed set of questions and expectations for the answer categorisation is represented in advance, the questions may be open ended, and the respondents may phrase their answers according to their understanding of the researcher's questions.

Semi-structured Interview

In a semi-structured interview setting, as observed by Karatsareas (2022), an interviewer maintains a list of open-ended questions and topics that should be covered during the interview. This form of an interview is advantageous as it allows openness and flexibility. Additionally, there is also a possibility to amend the questions partway through the interview (Ruslin et al., 2022). The aim is to gain rich personal accounts of individuals' experiences of particular issues. This technique can be utilised in various research areas. However, the interviewee's willingness to talk is an essential tool. In interviews, requiring sensitive or personal information as an ethical issue may impact the research project. The interviewer's ability to gain and keep the interviewee's trust is crucial. A semi-structured interview technique is widely used in the social sciences (Mannan & Afni, 2020). It offers the flexibility of conducting both party-driven interviews and ethnographic data collection and can be used when the respondent cannot write. Semi-structured interviews produce rich narrative information and can help us understand how an individual's understanding shapes practice and how more than one person experiences a particular issue. An additional advantage of these interviews is that, unlike structured interviews or questionnaires, their format allows for displaying emotions and creating a close and trusting interpersonal relationship (Adeoye-Olatunde & Olenik, 2021). A semi-structured interview is a meeting where the interviewer does not strictly follow a formalised list of questions. They will ask more open-ended questions, allowing for a discussion with the interviewee rather than a straightforward question-and-answer format. They will go into more depth about a particular area if the interviewee's response requires further elucidation. Naz et al. (2022, p. 4) defined semi-structured interviews as "essentially researcher-led in that the interviewer will have an interview guide specifying the topics to be covered but is prepared to help the interviewee to go beyond the items listed and does not stick rigidly to the sequence in which they appear". Semi-structured interviews are described as a flexible tool that allows interviewers to focus on a specific discussion while allowing adaptability (Osborne & Grant-Smith, 2021). A semi-structured interview guide typically includes critical and potential follow-up questions to explore a topic thoroughly. This format allows the interviewer to examine responses in greater depth while focusing on the conversation's central theme. It provides the flexibility to probe for more detailed insights while ensuring that the discussion remains aligned with the key objectives of the interview.

Unstructured Interview

Studies use the unstructured interview approach primarily because the approach helps the researcher to explore perceptions carefully within the context of the interviewee's experiences (Mashingaidze, 2021). Hence, it enables a detailed understanding of the research phenomenon. For example, since detailed literature regarding the operation of the Forex Bureau utilised in developing countries is lacking, the researcher employs an unstructured interview because the approach tolerates direct involvement with the informants through a personification process and through a one-to-one encounter during the interview that is vital to interact and gain a detailed understanding of the research phenomenon. Furthermore, the approach demands a few challenging decisions (Hammad, 2023). For instance, under the approach by James (2021), the researcher needed a more significant number of general questions to ask, which may also risk a loss of specific data. In the case of the research, the approach required the researcher to possess an in-depth understanding of the operation of the Forex Bureau to ask questions. However, as the literature in the area was limited, it was particularly challenging to obtain such questions. Moreover, the researcher must carefully listen to the informants during the interview to gain specific information. However, the informants might provide specific information as they see fit. It is also essential that the researcher be aware of any leading questions. During the interview, the researcher needed to thoroughly understand the Forex Bureau to establish a strong rapport and effectively engage with the informants. This knowledge helped the researcher connect with the participants, ensuring meaningful discussions and allowing the informants to share their expertise confidently. It is worth noting that the approach involved several stages that are discussed in subsections to follow shortly. Regarding the research objective and questions, the approach was suitable for a detailed exploration of the operation of the Forex Bureau. An unstructured interview is generally used to ensure direct involvement with the informants (Hasni & Mubarak, 2021). The interview is conducted by asking open-ended questions and allowing the informant to express their feelings and experiences. This approach requires a minimum level of predetermined questions as the researcher is flexible enough to raise general questions in response to the ever-emerging data without predetermined questions. This approach allows a researcher to understand the research phenomenon through the informants' perceptions of their experiences and views.

This study adopted a structured interview approach as the structured interview format standardises the interviewing process using consistent interview questions, behavioural role-playing, and demonstrative methods (Rashidi et al., 2024; Shea and

Barney, 2015). This leads to higher levels of reliability of the measures obtained through the structured interview process than the unstructured interview method. In an unstructured interview, interviewers are free to ask questions they deem relevant and can decide when to ask them, provide responses, and seek clarification or positive reinforcement, among other things. Dana et al. (2013) and Van der Zee et al. (2002) were confident that structured interviews yielded good data and, as noted already, allowed specific questions to be asked. In addition, when researchers are attempting to assess variation across several categories (cross-category comparisons), unstructured interviews "pose insurmountable problems of reliability and replicability" because the interviewers have to make choices about whether a particular benchmark has been met, and this is in effect trial and error at a local level. Even concerning the Mannion et al. (2009), p. 4) study on cancer, they "found it difficult to ensure that their studies produced comparable results", with relatively consistent results across categories but "different results for different types of animals." Finally, Bihu (2020) stated that unstructured interviews generated excessive data and occasionally confounded the agent categories of interest to the researchers (agents versus agencies in their case) and that this problem is common. Semi-structured interviews, usually seen as less structured than structured interviews, have the strengths of both structured and unstructured interviews (Gibson & Abrams, 2003). That is, they allow for at least some standardised data to be derived from the interviews while also allowing the possibility of obtaining some informants' free-range thoughts as part of the interview. However, semi-structured is less structured, only in terms of rules guiding the use of questions. The interview process should still be systematic. Hence, they can also have some drawbacks of less-structured methods, such as an inherent waste of time if specific questions are never asked or very unclear wording that results in varied interpretations of the question across interviews. Those problems, in turn, can result in a mix of structured-like data that is generalisable and more unstructured data that is highly specific to the given respondent interview. Therefore, it is said that "semi-structured interviews can be viewed as a compromise because they balance the structure of predefined questions with the flexibility to explore topics in more depth based on the interviewee's responses".

7.2 Selection of Interview Expert Panel

The maximum number of interviews in most qualitative studies can be relatively small compared to the sample size of quantitative research. This can challenge many researchers owing to the difficulty of interviewing qualified experts. Purposeful sampling has its roots in solid ethnography, sociology, and anthropology. Interview participants are selected because they experience specific phenomena, are involved in a particular situation, or can provide a meaningful understanding to the researcher. This requires that one select interview participants in such a way that the selection aligns with the research objectives. The person conducting the interview is a vital part of this process. When conducting structured interviews as part of the processes of phenomenology or grounded theory, it is essential to establish a set of rules called an interview taxonomy. It offers the methodological conduct of the interview process to prevent the destruction of the reported data and result in multiple interviews. It is vital for the success of the research. It is used as a resource for researcher's expertise. This involves presuppositions determined by the researcher and the research. The identification and selection of potential interview experts are essential in qualitative studies.

Nancy (2003) solicits the inclusion of key experts in the area of concern to solicit the latest thinking and give directions for policymaking. He also recommended the integration of experts' opinions with existing ones to formulate a research base for policymaking. However, the potential for bias is very likely when selecting experts for the panel. Rao and Perry (2003) emphasise the need for neutrality and impartiality when selecting the experts to make up the interview panel. However, this issue may not be completely resolved, as experts can sometimes display tendencies toward misunderstanding. A range of explicit selection criteria should be adopted and outlined as a basis for evaluating results and establishing the study's potential importance regarding other setups and populations since panel experts serve as the cornerstone of the interview process (Yeong et al., 2018). Avella (2016) opined that several criteria can be considered in the selection process for panel experts. Notably, two recommendations were mentioned. Firstly, the experts need to exhibit a sound knowledge base of the issue under review. Secondly, the experts should serve as professional representatives for the population's seamless acceptance of their suggestions. In a similar projection, Pelcovitz et al. (1997) advocate four requirements to be met by panel experts in an interview, namely experience in and knowledge of the subject of concern, willingness and capacity for participation; enough time for the entire process; and coherent and effective communication skills. For this study, the selection of experts for the panel of the interview technique was carried out in conformity with the attainment of at least five (5) of these criteria:

1. Knowledge: The individuals must have sufficient knowledge in the field of information and communication technology, knowledge in the field of modular construction, and knowledge in construction management.
2. Academic qualification: The individuals must have acquired a degree (National Diploma, Bachelor's, Master's, Ph.D.) in construction management.
3. Experience: The individuals should have a background of working in the construction industry or being involved in innovative building techniques projects. This can be with consultancy organisations, contracting organisations or government establishments.
4. Employment: The individuals must be currently serving or have served as professionals in any institution, agency, department, business, organisation, or company.
5. Authorship: They must h authored or co-authored research publications in peer-reviewed journals in construction management or facilities management; given presentations at conferences, symposiums, professional meetings, or workshops.

6. Recognition: They must have served as a peer-reviewer for at least one manuscript from the editor of a revered journal with an emphasis on construction methodology.
7. Teaching: They may have carried out teaching lessons or classes, notably focusing on construction methodology, and given presentations at workshops that aim to proffer professional advisory or technical expertise in construction methodology.
8. Research: They must have been engaged in research functions or duties, or must have submitted at least one research proposal, or have obtained a research grant or funding from a funding agency (government/private) in the area of innovative buildings or alternative building methods.
9. Professional membership: They must be a member of a professional body (most notably in construction or the built environment).
10. Willingness: They must be willing to participate in the interview process.

Astringent measures were adopted for this study, with the choice of five criteria to be met by the selected experts. This is differs from the prescribed criteria adopted by Muis et al. (2014) and Carrera and Mack (2010), who opted for two criteria: work experience and involvement in the particular nature of the project. Moreover, the criteria chosen for this study were in tandem with the prescriptions by Adler and Ziglio (1996), with the notable inclusion of the residency of the expert. This was considered necessary because experts need to base their opinions on their communal experience of facilities management and its cultural context. Furthermore, there was a need to set the minimum criteria for this study at five because the process may be undermined if the experts selected do not possess a proven track record in the area of concern, requisite qualifications, and specialist knowledge (Terwee et al., 2018).

The expert panel was recruited through emails that contained a general overview of the study. Also, their consent for inclusion in the interview was solicited, and those whose participation was confirmed were sent a comprehensive description of the interview process. Furthermore, a request was made for the confirmed experts' curriculum vitae to ascertain their areas of expertise and assess their conformity with the stipulated qualifying criteria. After verifying their curriculum vitae, it was affirmed that all the experts selected for the study met the qualifying criteria. The study identified 17 professional experts with basic modularity methodology knowledge and hands-on experience with modular construction and alternative building practices. These professionals practise in the built environment, academia, and financial technology sector in South Africa. Of the 17 prospective interviewees identified and contacted, only 12 were available, representing a 71% response rate. According to Hansen (2022) and Braun and Clarke (2023), the saturation point for interview-based data collection is between seven and 12 interviews, which justifies the adoption of the 12 interviewees in this study.

7.3 Interview Framework and Approach

Interviews often generate data that is both rich and abundant, leading to challenges in separating, interpreting, summarising, and reporting this information. These tasks can be highly technical and complex (Sivakumar & Susila, 2021). However, Xia et al. (2024) opined that a self-structured interview can be held under some degree of control, which means clarity and acceptable secure understanding. Through purposive sampling, participants are chosen to ensure diversity and representativeness, enhancing the richness and depth of the data collected. Subsequently, an interview protocol is developed, outlining the structure, questions, and topics to be covered during the interviews (Palinkas et al., 2015). This procedure finds a compromise between giving flexibility for investigation and follow-up and offering a structure for uniformity across interviews. Informed permission, which emphasises voluntariness, confidentiality, and the study's goal, is sought from participants before the start of interviews. Afterwards, the interviews are held in an atmosphere of mutual respect and candour, which enables participants to express their opinions and experiences openly. Probing and active listening strategies are used to obtain in-depth answers and delve into the subtleties of participants' narratives.

Data is documented throughout the interview using either audio recording, note-taking, or a mix of both methods to ensure accuracy and fidelity for subsequent analysis. Transcribing audio or video recordings into written text for analysis after data gathering could be required. The thematic or content analysis finds data patterns, themes, and insights. A more comprehensive knowledge of the study is achieved by interpreting the results in light of the research objectives, contextual circumstances, and body of current literature. This study used the structured interview approach, as previously mentioned. The interview was conducted using the Microsoft Teams online platform adopted by Chia et al. (2020) and Frey and Bloch (2023). This platform was chosen because it provides a comprehensive structure for interviewing study participants while taking essential factors such as accessibility, communication, data security, collaboration, and technological dependability into account. It is a good option for researchers who want to conduct interviews remotely while upholding methodological rigour and data integrity because of its integrated features and user-friendly interface.

7.4 Interview Specific Objectives

The review of extant literature revealed the various dimensions and attributes related to technology adoption in different fields of human endeavour. Also, several theories and models have been postulated to inform the adequateness and preparedness of adopting various technological innovations over the years. However, after exhaustively reviewing literature that dwelt on the theories above and models for modular construction adoption, there appear not to be any existential studies related to modularity methodology for the construction industry with South Africa as the geographical base. It is important to note that modularity methodology serves as one of the conveyors of the 4IR. Hence, such sophisticated construction methodology innovation is expected to be very unlikely in this part of the world. Concerning theoretical gaps mentioned above, considering the importance of construction methodology to the built environment, it is pertinent to develop a gateway that will as serve as a base for modularity methodology for the construction industry in South Africa. The interview technique helped test the attributes identified from the literature, which were considered significant in the drive to adopt modularity methodology for the construction industry. From the opinions of the interview experts for the study, the main and sub-attributes considered to propagate the adoption were ascertained based on their significance. However, the aim was to develop a conceptual model, which would be subsequently validated. Having established that the aim of engaging in the interview is to validate a conceptual model which is attributed to modularity methodology for the construction industry, the following were set as the specific objectives for the interview:

ISO1 To identify the main attributes and sub-attributes for modularity methodology in the construction industry and to evaluate whether these attributes would aid the adoption.

ISO2 To determine the influence of reassembly on modularity methodology for the construction industry.

ISO3 To determine the influence of maintenance on modularity methodology for the construction industry.

ISO4 To determine the influence of cultural perception on modularity methodology for the construction industry.

ISO5 To determine the extent to which modularity methodology can be adopted for the different types of buildings, and

ISO6 To predict the extent to which the various outcomes of modularity methodology can serve as drivers for adopting the construction method.

7.5 Analysis of Data from Interview

The tool used for analysing the interview result, "Atlas. ti", is a cutting-edge qualitative data analysis software (Gulsia & Yadav, 2023). It offers a variety of text mining and visualisation tools to excel in the behind-the-scenes technical aspects of good qualitative research. Okoliko and Wildschut (2024) advocate for coding text segments and applying these codes across multiple datasets. This approach enhances the analysis by allowing researchers to identify patterns and themes across different qualitative data consistently. It supports quantitative data analysis functions such as hierarchical clustering and factor analysis. It delivers platform-specific word cloud analysis with easier integration with data, analysis of keywords, thematic analysis and unmatched dictionary capability. Apart from minimising data loss and increasing access to stored data, Atlas ti provides efficient data management, which is crucial in research investigations (Gupta et al., 2024). Atlas ti and other data analysis software provide dual environment platforms (independent text coding for qualitative data). Zhang et al. (2023) opined that Atlas ti allows the automatic replication and increasing robustness of researchers' capacity and units of analysis, thus reducing the coding process and redundancy through inter-coder reliability, for example. In addition, Atlas ti's strengths include adding a series of parallelising phases, procedures, and mechanisms meant for optimising the management and reproducibility of analysed data while maintaining informant confidentiality (Al-Shuwaili et al., 2023; Rahmani et al., 2022). This is done by supporting independent coding or inter-method coding experimentation, with reduced costs in time and specific investments.

Smit and Scherman (2021) opined that Atlas ti is a comprehensive software that can analyse qualitative data. Therefore, the software is multipurpose in its application. One of Atlas ti's main advantages is its coding facility with text search and highlight extraction (Soratto et al., 2020). The facility could contribute mainly to coding interviews where a large amount of qualitative data is obtained, and analysis is time-consuming and complex. In order to code the data does not mean to be simple in its layout. Using the user-friendly screen, codes could be evaluated using different items. The stored exam items could also be used to conduct robust analysis. In creating complex codes, cross-breeding different cycles and using binary items would contribute to analysing categories using advanced items such as coding clusters and advanced exploration tools. Atlas ti has created an option to move fast utilising low resources by transferring the elaborate files and the ability to import files to other researchers in a network setup or to support data evaluations (Gonzalez-Prieto et al., 2020). This is a software which offers a powerful toolkit and also a user-friendly environment to facilitate qualitative data analysis. The software is often used in social sciences, business, health-related research, and other fields to evaluate various types of data. The strength of Atlas ti comes from its capability to work with large data sets, make connections, identify themes, and analyse qualitative data in an advanced and effective manner (Choe et al., 2022). The software package possesses innovative features for small sample research and can be used in various other mediums. For example, the relatively new feature of multimedia research is that it can analyse almost all types of data, including text, visual, and sound.

7.6 Ensuring Validity and Reliability of Insights

One viable way of determining sound and dependable results from the interview process is by ascertaining the consistency and viability of the data used. However, Belotto (2018) noted that ascertaining the consistency and viability of the interview process is a significant challenge for the interviewer. For clarity, Downing (2004, p. 1006) stated that "consistency refers to the reliability and stability of assessment outcomes". Considering that an interview falls under the categorisation of the qualitative research method, it is attributed to its challenges in establishing statistical consistency (O'Connor & Joffe, 2020). This results from the high possibility that the different panels will reach a different conclusion, subject to their level of knowledge in the area under focus (Zohrabi, 2013). Hence, to ensure consistency in the process, it is pertinent that the process is transparent to the experts while also ensuring a basic understanding of the purpose of the process (Dorussen et al., 2005). For this study, to ensure the consistency and validity of the interview process, a clear and detailed explanation of the aim of the process was spelt out to the experts. Firstly, a clear explanation of the rudiments of modularity methodology was provided. Also, the procedural steps for answering the questions were indicated so as not to misunderstand what needed to be said by the panellists. Furthermore, to entrench consistency and credibility in the interview process, the experts' opinions were ensured to be in line with truthfulness, credibility, conformity, and response consistency. Also, selecting experts is a proven basis for attaining reliability and validity in the interview process (O'Connor & Joffe, 2020). Therefore, as stated previously, proper care was taken to ensure that the panel's selected experts were proven individuals with adequate knowledge of and experience in the subject area. Moreover, the anonymity of the experts was ensured, eliminating the 'bandwagon' effect, thereby consolidating the strengths of the interview method. This also ensures the internal validity of the outcomes since no expert is being coerced or under pressure while giving opinions on questions.

7.7 Summary

This chapter validated the proposed conceptual framework through expert interviews, ensuring its relevance and applicability in the construction industry. The process involved careful selection of experts, structured interview design, and thematic analysis of qualitative data. Measures to ensure validity and reliability strengthened the credibility of findings. The results confirmed the framework's robustness while identifying refinements, positioning it as a practical and adaptable tool for promoting modularity methodology.

References

Adeoye-Olatunde, O. A., & Olenik, N. L. (2021). Research and scholarly methods: Semi-structured interviews. *Journal of the American College of Clinical Pharmacy, 4*(10), 1358–1367.

Adler, M., & Ziglio, E. (1996). *Gazing into the oracle: The Delphi method and its application to social policy and public health.* Jessica Kingsley Publishers.

Al-Shuwaili, H., Kakhki, M. S., Al-Aridhee, S., Ghorbani, A., Entezari, S., & Sardarabadi, M. (2023). Performance enhancement of triplex-tube heat storage unit using branched fins during solidification and melting processes: A 2D numerical parametric investigation. *Thermal Science and Engineering Progress, 38*, Article 101653.

Avella, J. R. (2016). Delphi panels: Research design, procedures, advantages, and challenges. *International Journal of Doctoral Studies, 11*, 305.

Belotto, M. J. (2018). Data analysis methods for qualitative research: Managing the challenges of coding, interrater reliability, and thematic analysis. *The Qualitative Report, 23*(11), 2622–2633.

Bihu, R. (2020). Using unstructured interviews in educational and social science research: The process, opportunity and difficulty. *Global Scientific Journals, GSJ, 8*(10). Available at SSRN: https://ssrn.com/abstract=4435828

Braun, V., & Clarke, V. (2023). Toward good practice in thematic analysis: Avoiding common problems and be (com) ing a knowing researcher. *International Journal of Transgender Health, 24*(1), 1–6.

Busetto, L., Wick, W., & Gumbinger, C. (2020). How to use and assess qualitative research methods. *Neurological Research and Practice, 2*(1), 14.

Carrera, D. G., & Mack, A. (2010). Sustainability assessment of energy technologies via social indicators: Results of a survey among European energy experts. *Energy Policy, 38*(2), 1030–1039.

Chia, C. K., Ghavifekr, S., & Razak, A. Z. A. (2020). Online interview tools for qualitative data collection during Covid-19 pandemic: Review of web conferencing platforms' functionality. *Malaysian Journal of Qualitative Research, 7*(1), 95–106.

Choe, Y., Lee, J., & Lee, G. (2022). Exploring values via the innovative application of social media with parks amid COVID-19: A qualitative content analysis of text and images using ATLAS. ti. *Sustainability, 14*(20), 13026.

Dana, J., Dawes, R., & Peterson, N. (2013). Belief in the unstructured interview: The persistence of an illusion. *Judgment and Decision Making, 8*(5), 512–520. https://doi.org/10.1017/S1930297500003612

Dorussen, H., Lenz, H., & Blavoukos, S. (2005). Assessing the reliability and validity of expert interviews. *European Union Politics, 6*(3), 315–337.

Frey, T. K., & Bloch, B. S. (2023). Using Microsoft Teams to facilitate asynchronous online focus groups. *International Journal of Qualitative Methods*. https://doi.org/10.1177/16094069231211251

Galana, H. N., Degago, A. T., Tsegaye, A. G., & Endashaw, A. A. (2023). Factors affecting the quality and effectiveness of student teachers during their practicum experiences: The case of some selected colleges in Oromia, Ethiopia. *Asian-Pacific Journal of Second and Foreign Language Education, 8*(1), 39.

Gibson, P., & Abrams, L. (2003). Racial difference in engaging, recruiting, and interviewing African American women in qualitative research. *Qualitative Social Work, 2*(4), 457–476.

González-Prieto, Á., Perez, J., Diaz, J., & López-Fernández, D. (2020). Reliability in software engineering qualitative research through inter-coder agreement: A guide using Krippendorff's α & Atlas. ti. arXiv preprint arXiv:2008.00977

Gulsia, O., & Yadav, S. (2023). Conceptualizing network approaches for a successful farm entrepreneurship using ATLAS. ti®. *Sustainable Technology and Entrepreneurship, 2*(1), 100026.

Gupta, R., Nair, K., Mishra, M., Ibrahim, B., & Bhardwaj, S. (2024). Adoption and impacts of generative artificial intelligence: Theoretical underpinnings and research agenda. *International Journal of Information Management Data Insights, 4*(1), Article 100232.

Hammad, A. (2023). *Transformational leadership and employee innovative behaviour in the Egyptian travel agents: A mediation–moderation model*. Thesis. University of Plymouth. Available at: https://doi.org/10.24382/5083

Hansen, M. K., Midtgaard, J., Hegaard, H. K., Broberg, L., & de Wolff, M. G. (2022). Monitored but not sufficiently guided—A qualitative descriptive interview study of maternity care experiences and needs in women with chronic medical conditions. *Midwifery, 104*, Article 103167.

Hasni, A., & Mubarak, R. H. (2021). *Constraints of non-oil-related SMEs' access to bank funding: Credit discouragement incidents in oil-based economy* (Doctoral dissertation, University of Glasgow).

Hockey, J., & Forsey, M. (2012). Ethnography is not participant observation: Reflections on the interview as participatory qualitative research. In J. Skinner (Ed.), *The interview* (1st ed., p. 19). Routledge, Taylor & Francis. https://doi.org/10.4324/9781003087014

Hosseini, S., Khandakar, A., Chowdhury, M. E., Ayari, M. A., Rahman, T., Chowdhury, M. H., & Vaferi, B. (2022). Novel and robust machine learning approach for estimating the fouling factor in heat exchangers. *Energy Reports, 8*, 8767–8776.

Islam, M. A., & Aldaihani, F. M. F. (2022). Justification for adopting qualitative research method, research approaches, sampling strategy, sample size, interview method, saturation, and data analysis. *Journal of International Business and Management, 5*(1), 01–11.

James, J. E. (2021). Black feminist epistemological methodology: Bridging theory and methods to research health and illness. In *Black feminist sociology* (pp. 207–216). Routledge.

Karatsareas, P. (2022). Semi-structured interviews. *Research Methods in Language Attitudes, 99.*

Mannan, S., & Afni, M. (2020). Best practices of semi-structured interview method. *Chittagong Port Authority*, 1–12.

Mannion, R., Konteh, F. H., & Davies, H. T. O. (2009). Assessing organisational culture for quality and safety improvement: A national survey of tools and tool use. *BMJ Quality & Safety, 18*(2), 153–156.

Martin, D. A., & Jamieson-Proctor, R. (2022). Pre-service teachers' perceptions of problem-based learning for developing their mathematics teaching pedagogy. *Interdisciplinary Journal of Problem-Based Learning, 16*(1).

Mashingaidze, T. M. (2021). Zimbabwe's paradoxical 'new dispensation': Political harm, endemic impunity and unending silences, 2017–2020. *Revista Brasileira de Estudos Africanos, 6*(11).

Muis, K. R., Duffy, M. C., Trevors, G., Ranellucci, J., & Foy, M. (2014). What were they thinking? Using cognitive interviewing to examine the validity of self-reported epistemic beliefs. *International Education Research, 2*(1), 17–32.

Murto, P., Hyysalo, S., Juntunen, J. K., & Jalas, M. (2020). Capturing the micro-level of intermediation in transitions: Comparing ethnographic and interview methods. *Environmental Innovation and Societal Transitions, 36*, 406–417. https://doi.org/10.1016/j.eist.2020.01.004

Nancy, J. L. (2003). "Our world" an interview. *Angelaki, 8*(2), 43–54.

Naz, F., Agrawal, R., Kumar, A., Gunasekaran, A., Majumdar, A., & Luthra, S. (2022). Reviewing the applications of artificial intelligence in sustainable supply chains: Exploring research propositions for future directions. *Business Strategy and the Environment, 31*(5), 2400–2423.

O'Connor, C., & Joffe, H. (2020). Intercoder reliability in qualitative research: Debates and practical guidelines. *International Journal of Qualitative Methods, 19*, 1609406919899220.

Osborne, N., & Grant-Smith, D. (2021). In-depth interviewing. In *Methods in urban analysis* (pp. 105–125). Springer.

Palinkas, L. A., Horwitz, S. M., Green, C. A., Wisdom, J. P., Duan, N., & Hoagwood, K. (2015). Purposeful sampling for qualitative data collection and analysis in mixed method implementation research. *Administration and Policy in Mental Health and Mental Health Services Research, 42*, 533–544. https://doi.org/10.1007/s10488-013-0528-y

Pelcovitz, D., Van der Kolk, B., Roth, S., Mandel, F., Kaplan, S., & Resick, P. (1997). Development of a criteria set and a structured interview for disorders of extreme stress (SIDES). *Journal of Traumatic Stress, 10*(1), 3–16.

Rahmani, K., Shavaleh, R., Forouhi, M., Disfani, H. F., Kamandi, M., Oskooi, R. K., Foogerdi, M., Soltani, M., Rahchamani, M., Mohaddespour, M., & Dianatinasab, M. (2022). The effectiveness of COVID-19 vaccines in reducing the incidence, hospitalization, and mortality from COVID-19: A systematic review and meta-analysis. *Frontiers in Public Health, 10*, Article 873596.

Rankin, J., Wahl, H., Venkataraman, A., & Olszanski, T. (2023). Radio pulsar beam geometry down to the 100-MHz band: 76 additional sources within the Arecibo Sky. *Monthly Notices of the Royal Astronomical Society, 519*(3), 3872–3878.

Rao, S., & Perry, C. (2003). Convergent interviewing to build a theory in under-researched areas: Principles and an example investigation of Internet usage in inter-firm relationships. *Qualitative Market Research: An International Journal, 6*(4), 236–247.

Rashidi, F. U., Mohsini, M. H., & Mega, B. (2024). A framework for security improvement on usage of mobile money application based on iris biometric authentication method. *Information Security Journal: A Global Perspective, 33*(6), 678–690. https://doi.org/10.1080/19393555.2024.2347240

Rinaldo, R., & Guhin, J. (2022). How and why interviews work: Ethnographic interviews and meso-level public culture. *Sociological Methods & Research, 51*(1), 34–67.

Røed, R. K., Baugerud, G. A., Hassan, S. Z., Sabet, S. S., Salehi, P., Powell, M. B., Riegler, M. A., Halvorsen, P., & Johnson, M. S. (2023). Enhancing questioning skills through child avatar chatbot training with feedback. *Frontiers in Psychology, 14*, 1198235.

Ruslin, R., Mashuri, S., Rasak, M. S. A., Alhabsyi, F., & Syam, H. (2022). Semi-structured Interview: A methodological reflection on the development of a qualitative research instrument in educational studies. *IOSR Journal of Research & Method in Education (IOSR-JRME), 12*(1), 22–29.

Scheinin, A., Kantonen, O., Alkire, M., Långsjö, J., Kallionpää, R. E., Kaisti, K., Radek, L., Johansson, J., Sandman, N., Nyman, M., Scheinin, M., Vahlberg, T., Revonsuo, A., Valli, K., & Scheinin, H. (2021). Foundations of human consciousness: Imaging the twilight zone. *Journal of Neuroscience, 41*(8), 1769–1778.

Shea, S. C., & Barney, C. (2015). Teaching clinical interviewing skills using role-playing: conveying empathy to performing a suicide assessment: A primer for individual role-playing and scripted group role-playing. *Psychiatric Clinics, 38*(1), 147–183.

Sivakumar, V. P., & Susila, C. (2021). Effectiveness of self-care measures on knowledge, self-efficacy and performance status among cancer patients. *Journal of Caring Sciences, 10*(1), 1. https://doi.org/10.34172/jcs.2021.003

Smit, B., & Scherman, V. (2021). Computer-assisted qualitative data analysis software for scoping reviews: A case of ATLAS. ti. *International Journal of Qualitative Methods, 20*, 16094069211019140.

Soratto, J., Pires, D. E. P. D., & Friese, S. (2020). Thematic content analysis using ATLAS. ti software: Potentialities for researches in health. *Revista Brasileira de Enfermagem, 73*, e20190250.

Terwee, C. B., Prinsen, C. A., Chiarotto, A., Westerman, M. J., Patrick, D. L., Alonso, J., Bouter, L. M., de Vet, H. C. W., & Mokkink, L. B. (2018). COSMIN methodology for evaluating the content validity of patient-reported outcome measures: A Delphi study. *Quality of Life Research, 27*(5), 1159–1170.

Thelwall, M., & Nevill, T. (2021). Is research with qualitative data more prevalent and impactful now? Interviews, case studies, focus groups and ethnographies. *Library & Information Science Research, 43*(2), Article 101094. https://doi.org/10.1016/j.lisr.2021.101094

Van der Zee, K. I., Bakker, A. B., & Bakker, P. (2002). Why are structured interviews so rarely used in personnel selection? *Journal of Applied Psychology, 87*(1), 176.

Xia, X., Indiran, L., Kohar, U. H. B. A., & Wang, Y. (2024). Understanding consumption motivation in confinement centres through in-depth interviews and nvivo analysis. *Educational Administration: Theory and Practice, 30*(5), 5119–5132.

Yeong, M. L., Ismail, R., Ismail, N. H., & Hamzah, M. (2018). Interview protocol refinement: Fine-tuning qualitative research interview questions for multi-racial populations in Malaysia. *Qualitative Report, 23*(11).

Yoo, K. M., Kim, J., Kim, H. J., Cho, H., Jo, H., Lee, S. W., Lee, S.-g. & Kim, T. (2022). Ground-truth labels matter: A deeper look into input-label demonstrations. arXiv preprint arXiv:2205.12685

Zhang, W., Knox, D., & Prabhakar, G. (2023). Risk in active sport tourism projects: Narratives from managers in the Chinese event industry. *Journal of China Tourism Research, 19*(2), 176–196.

Zohrabi, M. (2013). Mixed method research: Instruments, validity, reliability and reporting findings. *Theory and Practice in Language Studies, 3*(2), 254–262.

Chapter 8
Expert Opinions and Industry Insights Results

Abstract This chapter presents the findings derived from the interview sessions conducted with selected experts. The responses have been organised and reported in alignment with the research objectives established for this study, ensuring that the analysis directly addresses the core aims of the investigation. To achieve triangulation and enhance the credibility of the results, the study adopted a qualitative research methodology, with interviews serving as a key data collection method. The interviews were conducted with professionals from the built environment who possess expertise and practical knowledge of modularity methodology in the construction industry. Their insights were invaluable in providing both theoretical and practice-oriented perspectives on the subject. Following the data collection, all interview responses were carefully transcribed and systematically analysed. A content analysis approach was employed, supported using ATLAS.ti software, to identify recurring themes, patterns, and meaningful interpretations within the data. The outcomes of this analytical process are presented in this chapter, offering a structured account of the expert perspectives gathered. This chapter not only demonstrates the theoretical and practical applicability of modularity methodology but also lays the groundwork for future discussion and interpretation in later sections by aligning the findings with the study's goals.

8.1 Introduction

The study identified 17 professional experts with basic knowledge of construction modularity methodology. These professionals practise in South Africa in the built environment and academia. Out of the 17 prospective interviewees identified, only 12 were available, representing a 71% response rate. The demographic of the interviewee is documented in Table 8.1.

From the information in Table 8.1, the interviewees possess between five and 25 years of experience, with an average of 14 years. Seven (7) of the interviewees have less than 15 years of experience, while the remaining five have practised their

P. Adekunle et al., *Modularity Methodology*,
https://doi.org/10.1007/978-3-032-12918-5_8

Table 8.1 Interviewee's demographic information

Interviewee code	Sector	Years of experience	Highest academic qualification	Discipline/Role
P1	Built environment	25	MSc	Construction Project Manager
P2	Built environment	10	MSc	Research Specialist
P3	Academia	22	PhD	Civil Engineer
P4	Built environment	12	PhD	Urban and Regional Planner
P5	Academia	28	MSc	Construction Project Manager
P6	Built environment	5	MSc	Architect
P7	Built environment	20	BSc	Housing Developer
P8	Academia	8	PhD	Quantity Surveyor
P9	Built environment	12	MSc	Contractor
P10	Built environment	11	BSc	Construction Project Manager
P11	Built environment	5	BSc	Architect
P12	Academia	15	PhD	Quantity Surveyor
Mean		14.25		

profession for more than 15 years. Three of the interviewees possess only undergraduate degrees in their respective professions, five possess master's degrees, and four also possess doctoral degrees in their respective fields. A good number of them (P1, P2, P5 and P9) also possess a honours degree in health and safety practices. P2 submitted that:

> *"Construction projects always carry a number of risks to the health and safety of the personnel. Managers who have received advanced health and safety training are better able to identify hazards, put mitigation plans in place, and create a safer work environment."*

Considering the professional groups to which the interviewees belong, it can be summarised that four of the interviewees work in the academic space with a built environment background. The remaining eight are built environment professionals actively practising in the sector. Based on the information gathered from these professionals, it could be inferred that the interviewees are well-versed in both the built environment and academic sectors, with a wealth of years of experience in both sectors. Hence, their opinion on the interview sessions can be relied upon as a valid means of collecting valuable data for this study.

8.2 Thematic Analysis of Interviews

Having carried out a detailed systematic literature review, the study identified nine constructs (including the gap) that influence modularity methodology for the construction industry. The identified constructs were presented to the interviewees to validate each construct identified. Their opinions were recorded in the following subsections. In addition, the interviewees were questioned on the outcome of modularity methodology for the construction industry and the applicability of modularity methodology to different kinds of buildings.

8.2.1 Supply Chain

Interviewees were questioned about their opinions on the influence of supply chain attributes on modularity methodology for the construction industry.

Response:
P1 stated that…
"*…a well-optimised supply chain facilitates the timely delivery of modular components, crucial for project schedules…*"
P4 opined that…
"*…the coordination between supply chain and on-site construction teams is essential for seamless assembly and installation of modular units…*"
P5 remarked that….
"*…supply chain visibility and transparency enable better tracking of components and materials, ensuring they meet project specifications…*"
P6 added that…
"*…the collaboration between supply chain partners fosters innovation and the development of new modular construction techniques…*"

These viewpoints emphasise how crucial it is to have a productive and well-structured supply chain while using modularity methodology for construction. To keep the project on schedule, prefabricated components must be delivered on time. The project timeline may be significantly impacted, and expenses may increase if certain components are not received on time. Furthermore, efficient coordination and communication guarantee that modular components may be installed and assembled on-site promptly and accurately when they arrive. This collaboration promotes a more efficient workflow by preventing mistakes and delays during building.

Furthermore,
P2 stated that…
"*…supply chain disruptions can hinder the implementation of modularity methodology, leading to delays and cost overruns…*"
P3 opined that…

"...*collaborative relationships with suppliers are essential for sourcing high-quality modular components*..."

P7 remarked that....

"...*supplier performance monitoring is vital for maintaining high standards and ensuring that any issues are promptly addressed*..."

P8 added that...

"...*efficient warehousing and inventory management practices support the timely availability of modular components*..."

Supply chain disruptions, such as material delivery delays, shortages, or logistical problems, can severely impact the modular building process. The necessity for accelerated shipping, storage fees for idle modules, or fines for project completion delays might result from these disruptions, which can further raise expenses and cause project delays. Reducing these disruptions and keeping project budgets and schedules intact requires effective risk management techniques. These rules aid in maintaining an effective workflow and preventing bottlenecks by preventing lost or delayed items.

In addition,

P9 stated that...

"...*collaborative planning with suppliers ensures that modular components meet the specific design and quality standards required for construction projects*..."

P10 opined that...

"...*effective supply chain strategies can reduce waste and optimise the use of materials, supporting sustainable construction practices*..."

P11 remarked that....

"...*modular construction projects can benefit from a lean supply chain approach, focusing on minimising waste and maximising value*..."

P12 added that...

"...*local sourcing of materials can reduce transportation costs and lead times, benefiting projects employing modularity methodology*..."

Based on the participants' opinions, it can be deduced that working closely with suppliers during the planning phase is important to ensure that the prefabricated components used in modular construction align with the project's design specifications and quality standards. A lean supply chain focuses on eliminating non-value-adding activities and waste, maximising the overall value delivered to the project. This can lead to more efficient use of resources, reduced costs, and enhanced project performance.

8.2.2 *Productivity*

Interviewees were questioned about their opinions on the influence of productivity attributes on modularity methodology for the construction industry.

Response:
P1 stated that…
"…*modularity methodology in the construction industry significantly enhances productivity by streamlining the construction process*…"
P5 opined that…
"…*modularity methodology supports the parallel development of different project phases, significantly shortening the construction schedule*…"
P6 remarked that…
"…*the assembly line approach of modular construction increases the speed of module production compared to traditional methods*…"
P9 added that…
"…*controlled environment of modular fabrication allows for better quality assurance processes, leading to fewer delays and increased productivity*"
"…*modular construction enables faster project turnover, allowing firms to take on more projects*…"

These perspectives highlight how modular construction techniques increase overall productivity by streamlining the construction process. Compared to conventional on-site building, the modular approach allows for more controlled and efficient production because components are prefabricated in a factory. As a result, the process becomes quicker and more efficient, reducing the time and resources required to finish tasks. Assembly line procedures can be used in a factory to standardise and optimise the production of building modules. This tactic increases production speed while enhancing consistency and efficiency by segmenting tasks into discrete, repeatable processes.

Furthermore,
P2 stated that…
"…*quality control in modular construction leads to fewer on-site reworks, thus boosting productivity*…"
P3 opined that…
"…*modular construction allows for better resource allocation, as materials and labour can be efficiently managed in a controlled environment*…"
P8 remarked that…
"…*the ability to fabricate modules in parallel with site work results in a more streamlined and productive construction process*…"
P12 added that…
"…*modular construction's focus on lean manufacturing principles reduces inefficiencies and boosts overall productivity*…"

Modular construction's rigorous quality control processes guarantee that parts are produced in a regulated manufacturing environment to exacting specifications. This lessens the possibility of mistakes or flaws that must be fixed after the modules are assembled at the site. As a result, reduced on-site reworks save money and time while boosting output. Furthermore, in contrast to conventional building techniques, this regulated environment enables accurate resource management and optimisation, reducing waste and guaranteeing that labour is employed more efficiently.

In addition,
P4 stated that…
"*…modularity methodology supports the parallel development of different project phases, significantly shortening the construction schedule…*"
P7 opined that…
"*…productivity is enhanced by the ability to standardise and replicate successful module designs across multiple projects…*"
P8 remarked that…
"*…the ability to fabricate modules in parallel with site work results in a more streamlined and productive construction process…*"
P10 added that…
"*…enhanced predictability in modular construction schedules allows for better planning and resource management…*"

The perspectives expressed by the participants indicate that modular construction facilitates the concurrent development of several project phases. With modular construction, modules can be manufactured off-site. At the same time, site preparation and other tasks are being done, unlike traditional construction, where each phase frequently needs to be finished before the next can start. This simultaneous processing greatly shortens the total manufacturing time. A module design can be duplicated across multiple projects if it succeeds, lowering new design requirements and speeding up production. Because tested designs are utilised, this efficiency results in quicker project completion times and consistent quality.

8.2.3 Digitalisation

Interviewees were questioned about their opinions on the influence of digitalisation attributes on modularity methodology for the construction industry.

Response:
P2 stated that…
"*…the use of 3D modelling software allows for the detailed design and testing of modular units before physical construction begins…*"
"*…construction projects benefit from the real-time tracking and management capabilities provided by digital tools…*"
P5 opined that…
"*…data analytics tools provide valuable insights into project performance, helping to identify and mitigate risks…*"
P6 remarked that….
"*…enhanced precision in design and construction reduces the likelihood of errors and rework…*"

P12 added that…

"…*the integration of Internet of Things devices allows for real-time monitoring of construction site conditions*…"

These claims emphasise how crucial cutting-edge digital tools are to modular building. Using 3D modelling software, engineers and architects may produce accurate and comprehensive virtual representations of modular units. This makes it possible to test and optimise designs thoroughly in a virtual setting before starting construction. These tools enable the identification of patterns and trends in project performance by analysing vast amounts of data from multiple sources. Project managers can take proactive steps to limit potential risks, such as delays, budget overruns, or quality difficulties, by using these insights to identify them early.

Furthermore,

P3 stated that…

"…*automated scheduling and project management software helps optimise timelines and reduce delays*…"

P4 opined that…

"…*virtual and augmented reality technologies enhance stakeholder engagement and decision-making processes*…"

P7 remarked that….

"…*the ability to simulate and test modular designs digitally reduces the need for physical prototypes*…"

P9 added that…

"…*digital platforms allow for the efficient management of documentation and compliance records*…"

The participants recommended using project management software and automated scheduling to help plan and oversee building projects. Through the creation of exact timetables, effective resource allocation, and real-time progress tracking, these technologies expedite the planning and execution stages. To be more specific, before construction projects are created, stakeholders may envisage and engage with them thanks to the immersive and interactive environments that virtual and augmented reality (VR and AR) technology provides. These tools can facilitate presenting project plans, performing virtual walkthroughs, and seeing any problems early on.

In addition,

P1 stated that…

"…*through digitalisation, construction firms can easily integrate and coordinate various aspects of the modular construction process*…"

P8 opined that…

"…*digitalisation enables the collection and analysis of lifecycle data, supporting maintenance and facility management*…"

P10 remarked that…

"…*modularity methodology is accelerated by the ability to integrate and optimise processes digitally*…"

P11 added that…

"*...cloud-based platforms facilitate the storage and sharing of project data, ensuring accessibility and transparency...*"

The participants' viewpoints suggest that digital tools and technology facilitate the smooth integration and coordination of various components of the modular construction process for construction businesses. This implies that everything can be handled more effectively, including on-site assembly, design, manufacture, and logistics. Digitalisation makes gathering and analysing data thoroughly over a building's life-cycle possible. Proactive maintenance and effective facility management are made possible by using this data to track the state and functionality of building components.

8.2.4 Legal and Regulatory Considerations

Interviewees were questioned about their opinions on the influence of legal and regulatory consideration attributes on modularity methodology for the construction industry.

Response:
P1 stated that...
"*...compliance with building codes and standards is essential for the acceptance of modular construction practices...*"
P2 opined that...
"*...legal requirements for inspections and certifications can affect the timeline and cost of modular construction projects...*"
P4 remarked that...
"*...the construct documents must be clear with no ambiguity*", P4 "*Building codes may need to be updated to address the specific challenges and benefits of modular construction...*"
P5 added that......
"*...building codes may need to be updated to address the specific challenges and benefits of modular construction...*"

The participants underscored the importance of modular construction methods according to current building norms and requirements. Compliance guarantees that modular homes are dependable, safe, and up to the standards established by authorities in the field. Obtaining government clearance and expanding the modularity approach in the building sector depends on this acceptability. Legal requirements for certifications and inspections can greatly impact the cost and schedule of modular building projects. Modular construction necessitates rigorous inspection procedures at several stages since it involves both off-site manufacture and on-site assembly.

Furthermore,
P3 stated that...
"*...permitting processes must be streamlined to facilitate the timely approval of modular construction projects ...*"

P6 opined that…

"…*continuous review and adaptation of legal and regulatory frameworks are necessary to keep pace with advancements in modular construction technology and practices* …"

P7 remarked that…

"…*there should be policies stipulated for all classification of target groups involved in modular construction* …"

P12 added that…

"…*building warranty laws should cover modular buildings to ensure long-term quality and durability* …"

In order to expedite the timely approval of modular building projects, the participants highlighted the necessity of reduced permission procedures. Because modular construction requires exact timeframes and schedules, expedited permitting is essential. Permit delays can cause significant disruptions to the construction process, raising costs and delaying project completion dates. Legal and regulatory frameworks must be reviewed and modified regularly to keep up with the rapid improvements in modular construction technology and techniques. New technologies and creative methods in modular construction may render the current regulations antiquated or insufficient. Regular updates and changes are required to guarantee that these frameworks continue to be applicable and supportive of the most recent advancements in the industry and to advance safety, quality, and efficiency in modular construction.

In addition,

P8 stated that…

"…*governments can incentivise modular construction through tax benefits, grants, and other financial incentives*…"

P9 opined that…

"…*the government should formulate policy documents promoting modular construction*…"

P10 remarked that…

"…*labour laws and regulations must consider the different target groups, working conditions and practices in modular construction*…"

P11 added that…

"…*dispute resolution mechanisms should be in place to address conflicts specific to modular construction projects*…"

According to the participants, governments can significantly help promote modular building by providing financial incentives. These could include financial incentives to reduce the upfront costs and promote adoption, tax benefits for businesses that use modular techniques, and grants for the study and development of modular technology. Businesses would find modular building more financially feasible and appealing with such incentives. It is imperative to modify labour rules and regulations to account for the distinct features of modular construction. This entails considering the workforce's diversity, providing safe and suitable working conditions, and considering the unique procedures and techniques applied to modular projects.

8.2.5 Sustainability

Interviewees were questioned about their opinions on the influence of sustainability attributes on modularity methodology for the construction industry.

Response:
P1 stated that…
"…*sustainability significantly influences modularity methodology in the construction industry by promoting eco-friendly building practices*…"
P2 opined that…
"…*modular construction supports sustainability through efficient use of resources, optimising material usage, and reducing the carbon footprint of building projects*…"
P4 remarked that….
@@@@@ "…*energy efficiency is a key focus in sustainable modular buildings*…"
P5 added that…
"…*use of low-emission materials and finishes, improving indoor air quality and occupant health*…"

The participants emphasised the importance of sustainability in promoting the growing use of modularity approaches. This indicates that modular construction is becoming more appealing owing to the industry's shift to eco-friendly methods, such as cutting waste and minimising environmental effects. Sustainable construction practices are well-aligned with modular construction, which promotes ecologically friendly practices. Modular construction techniques contribute to the sustainability of building projects by maximising the quantity of materials used, reducing the overall carbon footprint, and using resources more effectively. This claim emphasises how modular building can reduce environmental effects and help manage resources.

Furthermore,
P3 stated that…
"…*the controlled factory environment in modular construction allows for better management of waste and recycling*…"
P6 opined that…
"…*minimises on-site disruptions and disturbances to the surrounding environment, preserving natural landscapes*…"
P7 remarked that…
"…*modularity allows for better integration of green technologies, such as solar panels and green roofs, into building designs*…"
P12 added that…

"...*modularity methodology in construction fosters innovation in sustainable building practices, setting new benchmarks for the industry*..."

Interviewees describe how modular construction's inherent factory-controlled environment improves waste management and minimisation capabilities. Materials may be cut more accurately in a factory, making recycling off-cuts more accessible. Compared to conventional on-site construction processes, this controlled approach produces less waste, promoting more environmentally friendly building practices. The early integration of green technologies, such as solar panels, green roofs, and energy-efficient systems, is made more accessible by modular architecture. Integrating these technologies into the building design enhances the construction project's overall sustainability.

In addition,
P8 stated that...
"...*modular construction supports sustainable urban development by enabling the rapid deployment of high-density housing*..."
P9 opined that...
"...*modular construction aligns with government policies and incentives to reduce greenhouse gas emissions and promote green building practices*..."
P10 remarked that...
"...*modular construction aligns with government policies and incentives to reduce greenhouse gas emissions and promote green building practice*..."
P11 added that...
"...*modular buildings often feature advanced building automation systems that optimise energy use and environmental performance*..."

The assertions highlight how modular construction promotes rapid assembly of dwelling units, which is beneficial for sustainable urban development. With this approach, the increasing need for high-density housing in cities may be effectively satisfied, lessening the burden on the existing resources and infrastructure and fostering more environmentally friendly living conditions. The data shows that government measures to reduce greenhouse gas emissions align with modular building. By utilising green building techniques, such as energy-efficient designs and sustainable materials, modular construction can benefit from laws and incentives intended to encourage environmental sustainability.

8.2.6 Procurement Method

Interviewees were questioned about their opinions on the influence of procurement method attributes on modularity methodology for the construction industry.

Response:
P1 stated that...

"...*the most suitable procurement approach to modularity methodology is turnkey*..."

P4 opined that...

"...*early contractor involvement, facilitated by certain procurement methods, allows for better planning and integration of modular units*..."

P10 remarked that...

"...*clear procurement guidelines and frameworks are essential for ensuring compliance and quality in modular construction*..."

P12 added that...

"...*the selection of suppliers through strategic procurement methods can ensure a consistent supply of high-quality modular components*..."

According to the participants, modular construction is best suited for the turnkey procurement strategy, in which a single contractor handles design and construction. Furthermore, it is crucial to include contractors early in the project using certain procurement techniques. Early engagement guarantees that modular units are smoothly integrated throughout the construction process by allowing contractors to participate in the planning and design phases. This proactive involvement ensures that the modular components align with the project requirements, increases efficiency, and helps spot possible problems early.

Furthermore,

P2 stated that...

"...*design-build procurement can enhance modularity by fostering early collaboration between designers and builder*..."

P3 opined that...

"...*the use of reverse auction procurement methods can drive down costs, making modular components more financially viable*..."

P5 remarked that...

"...*procurement methods that emphasise pre-construction planning can significantly reduce the risks associated with modular construction*..."

P7 added that...

"...*public–private partnerships can leverage modular construction to achieve faster project delivery and cost savings*..."

According to interviewees, design-build procurement encourages early and ongoing collaboration between designers and builders by having a single company handle both the design and building stages. This integrated approach is especially advantageous for modular construction because it makes it possible to integrate the design and building processes seamlessly. Early collaboration ensures that constructability is considered when designing modular components, which optimises production and assembly processes, lowers mistake rates, and improves overall efficiency and quality. In reverse auction procurement, vendors compete with one another to deliver the best deal on their products or services. Through the competitive bidding process, suppliers are encouraged to submit their best prices, which can result in significant cost savings.

In addition,
P6 stated that…
"…*the selection of suppliers through strategic procurement methods can ensure a consistent supply of high-quality modular components*…"
P8 opined that…
"…*building long-term relationships with suppliers through strategic procurement can ensure continuous improvement in modular component quality*…"
P9 remarked that…
"…*risk management in modular construction is enhanced through procurement methods that promote early risk identification and mitigation, a system that encourages effective construction management*…"
P11 added that…
"…*private financing procurement strategy can promote modularity methodology*…"

The claims emphasise how crucial it is to select suppliers using tactical procurement procedures properly. The integrity and effectiveness of modular construction projects depend on a consistent supply of high-quality modular components, which businesses may ensure by choosing the proper suppliers. The remarks also emphasise how important it is to establish lasting connections with suppliers. These connections can improve the quality of modular components, which will enhance the entire building process and outcomes through continued collaboration and strategic procurement strategies.

8.2.7 Reassembly

Interviewees were questioned about their opinions on the influence of reassembly attributes on modularity methodology for the construction industry.

Response:
P3 stated that…
"…*the ability to reassemble components can shorten construction timelines, as pre-assembled modules can be quickly put together and taken apart as needed*…"
P4 opined that…
"…*the process supports circular economy principles, where materials and components are reused in new projects, minimising the need for virgin resources*…"
P10 remarked that…
"…*the approach can enhance the aesthetic appeal of modular buildings because components can be replaced or upgraded to stay current with design trends*…"
P11 added that…
"…*the ability to reassemble components can facilitate compliance with changing building codes and regulations*…"

The participants stated that the modularity methodology emphasises the speed at which pre-assembled modules may be assembled and taken apart, resulting in efficient construction. This modularity methodology shortens building times compared to traditional methods because modules are made off-site and quickly installed on-site. Modular building adheres to the circular economy's tenets by recycling parts and resources from earlier projects, cutting waste and the need for new raw materials while fostering environmental sustainability.

Furthermore,
P1 stated that…
"…*emphasis should be placed on the ease of disassembling and reusing components*…"
P2 opined that…
"…*reassembly allows modular construction projects to be more adaptable, facilitating changes and upgrades without extensive demolition*…"
P9 remarked that…
"…*the ability to disassemble and reassemble modules makes it easier to meet changing client needs and project requirements*…"
P12 added that…
"…*the ability to reassemble components can facilitate compliance with changing building codes and regulations*…"

The respondents' replies emphasise the importance of developing modular components for easy disassembly and reuse. This method improves sustainability by cutting waste and facilitating the effective reuse of materials in subsequent projects. The modular structure offers increased flexibility due to its re-assemblable pieces. Because of its versatility, a structure can be updated or changed with little disruption, negating the need for significant demolition and replacement.

In addition,
P5 stated that…
"…*the selection of suppliers through strategic procurement methods can ensure a consistent supply of high-quality modular components*…"
P6 opined that…
"…*building long-term relationships with suppliers through strategic procurement can ensure continuous improvement in modular component quality*…"
P7 remarked that…
"…*risk management in modular construction is enhanced through procurement methods that promote early risk identification and mitigation, a system that encourages effective construction management*…"
P8 added that…
"…*private financing procurement strategy can promote modularity methodology*…"

The assertions highlight how strategic procurement entails carefully selecting suppliers according to their track record of providing high-quality supplies. By doing this, building projects can maintain a steady supply of modular parts, which is essential for fulfilling deadlines and guaranteeing the quality of the finished product. Consequently, building and maintaining enduring relationships with suppliers encourages cooperation and confidence. Suppliers are encouraged to improve the quality of their modular components continuously owing to this ongoing engagement, which is advantageous to both parties and ultimately enhances project outcomes.

8.2.8 Maintenance

Interviewees were questioned about their opinions on the influence of maintenance attributes on modularity methodology for the construction industry.

Response:
P5 stated that…
"…regular maintenance of modular components can extend their lifespan, ensuring durability and long-term performance of modular buildings…"
P6 opined that…
"…maintenance planning should be integrated into the initial design stage so that components are easily accessible and serviceable…"
P11 remarked that…
"…the modular approach reduces maintenance costs because faulty parts can be replaced individually without disrupting the entire structure…"
P12 added that…
"…the ability to maintain and upgrade modular systems ensures compliance with evolving safety and building standards…"

The participants indicated that modularity methodology emphasises efficient maintenance practices by allowing targeted repairs and replacements. This approach ensures durability while reducing the overall cost of ownership compared to traditional methods. The design of modular systems supports ease of access to critical components, enabling proactive maintenance and preventing extensive structural damage. Modular maintenance also contributes to sustainability by extending component lifecycles and minimising waste.

Furthermore,
P1 stated that…
"…maintenance schedules should be standardised to ensure consistency and predictability in the upkeep of modular components…"
P2 opined that…
"…ease of component replacement enhances building flexibility, as upgrades and modifications can be carried out without significant downtime…"
P7 remarked that….
"…digital tools such as Building Information Modelling (BIM) can support predictive maintenance by tracking component performance…"

P8 added that…
"…planned maintenance strategies improve user confidence in modular construction, making it a more attractive option for clients…"

The respondents' replies emphasise that effective maintenance planning ensures reliability, client satisfaction, and adaptability of modular buildings. The ability to replace or upgrade components without extensive disruption supports both functionality and aesthetic relevance. Furthermore, integrating digital technologies strengthens maintenance processes by enabling proactive interventions and reducing lifecycle costs.

In addition,
P3 stated that…
"…maintenance contracts with specialised service providers can ensure continuous support and quality assurance for modular projects…"
P4 opined that…….
"…collaboration with component manufacturers in long-term maintenance agreements ensures availability of spare parts and technical expertise…"
P9 remarked that….
"…risk management in modular projects is enhanced when maintenance responsibilities are clearly defined in procurement contracts…"
P10 added that…
"…investment in maintenance strategies safeguards asset value, protecting returns for both private investors and public sector clients…"

The assertions highlight that structured maintenance strategies enhance modularity methodology by ensuring long-term performance, cost savings, and compliance. Strong collaborations with suppliers and service providers guarantee spare parts availability and continuous technical support. Clear allocation of maintenance responsibilities also improves risk management and project reliability, making modular construction a sustainable and cost-effective alternative in the industry.

8.2.9 Cultural Perception

Interviewees were questioned about their opinions on the influence of cultural perception attributes on modularity methodology for the construction industry.

Response:
P1 stated that…
"*…cultural norms and values influence how stakeholders perceive and embrace innovative construction practices…*"
P5 opined that…
"*…in cultures with a strong emphasis on tradition and heritage like South Africa, there may be a reluctance to depart from established construction practices…*"
P6 remarked that…

"*…cultural attitudes towards risk and uncertainty can influence the willingness to invest in and experiment with modularity methodology…*"
P7 added that…
"*…cultural values related to community and social responsibility may drive interest in modular construction…*"

The participants stated that a culture's collective beliefs, practices, and values influence stakeholders' perceptions of innovative building methods. Modularity methodology in construction, for example, may be well received in societies where innovation is highly regarded and promoted. Different cultural perspectives may place a higher value on tradition and be more averse to change, which could impact how new approaches are viewed and accepted. Long-standing building practices, for instance, are an integral component of the cultural identity of South Africa. This means that the locals, who emphasise maintaining traditional methods, may oppose or be hesitant to embrace modular building.

Furthermore,
P2 stated that…
"*…there may be resistance to change due to cultural knowledge, hindering the adoption of new methodologies like modularity…*"
P4 opined that…
"*…there may be resistance to change due to cultural knowledge, hindering the adoption of new methodologies like modularity…*"
P9 remarked that…
"*…cultural stigmas or biases associated with certain construction methods may impact the perception…*"
P11 added that…
"*…hierarchy and authority within the construction industry may affect decision-making processes…*"

The interviewees' comments indicate that the construction industry's prevailing hierarchical structure and authoritative decision-making processes can notably impact the adoption of novel approaches. Higher-ranking decision-makers may be reluctant to embrace innovations such as modularity because they fear upsetting long-standing procedures or incurring unknown consequences. To overcome these obstacles, organisational leadership must recognise the advantages of modularity and ensure that decision-making procedures are adaptable to changing circumstances. Modern approaches can be viewed differently depending on cultural stigmas or prejudices associated with the construction methods. Owing to these biases, stakeholders may support or oppose strategies based more on past impressions than impartial evaluations of their efficacy.

In addition,
P3 stated that…
"*…cultures that prioritise collaboration and flexibility may be more open to adopting modularity methodology…*"
P8 opined that…

"*...cultural diversity within construction project teams can foster innovation...*"
P10 remarked that....
"*...cultural perceptions of aesthetics and architectural style may influence preferences...*"
P12 added that...
"*...traditional construction practices deeply rooted in culture may be perceived as more reliable and preferable...*"

These claims imply that organisational cultures that value flexibility and teamwork highly are probably more open to using novel techniques such as modular construction. The efficient use of modular construction principles necessitates a culture that values teamwork, adaptability, and willingness to change. This demonstrates how organisational culture influences attitudes towards creativity and novel ideas in building projects. Teams made up of people with different cultural origins have access to a variety of viewpoints, concepts, and approaches to problem-solving. Implementing novel building techniques such as modularity methodology might result from this diversity, as can imaginative solutions and cutting-edge methods. It emphasises the benefits of cultural diversity for project success and business innovation.

8.2.10 Outcome of Modularity Methodology

Interviewees were questioned about their opinion on the outcomes of modularity methodology for the construction industry.

Response:
P1 stated that...
"*...modularity methodology in the construction industry results in increased efficiency...*"
P2 opined that...
"*...modular construction allows for faster project completion than traditional methods...*"
P3 remarked that...
"*...using standardised modular components reduces construction time and labour costs...*"
P4 added that...
"*...modularity methodology improves project scheduling and resource allocation...*"

According to the participants, the capacity to hasten project completion is one of the primary results of the modularity methodology for the construction sector. Construction times can be greatly reduced by manufacturing modules off-site in a controlled environment while on-site preparations are underway. Compared to conventional building methods, this methodology minimises weather-related delays and on-site disruptions, resulting in speedier project delivery. It emphasised how employing standardised modular components increases efficiency. Standardisation shortens the time needed for building activities by enabling repeatable production procedures and predictable on-site assembly. Modular architecture also helps reduce labour costs by maximising output and minimising customisation, which lowers overall expenses.

Furthermore,
P5 stated that…
"…*modular construction enables greater flexibility in project design and customization*…"
P6 opined that…
"…*the outcome of modularity methodology includes improved quality control*…"
P7 remarked that…
"…modular buildings often exhibit higher levels of precision and accuracy in construction…"
P8 added that…
"…*the standardised nature of modular components ensures consistency and reliability*…"

These claims highlight the greater customisation and design freedom that modular construction provides for projects. Standardised modules or components that are easily customisable to match project needs are used to construct modular buildings. Compared to conventional building methods, this flexibility enables architects and designers to better modify designs to specific site conditions, client preferences, and functional requirements. Enhanced quality control is one of the benefits of employing modular building technology. Modules are created under stringent quality requirements and conditions because modular building facilities have a regulated production environment. This regulated procedure lowers errors, minimises unpredictability, and improves building quality. As a result, modular buildings often demonstrate higher levels of consistency and reliability regarding structural integrity and finish quality.

In addition,
P9 stated that…
"…*modular construction reduces waste and promotes recycling of materials*…"
P10 opined that…
"…*the outcome of modularity methodology includes cost savings for construction firms and clients*…"
P11 remarked that…
"…*modularity methodology improves safety on construction sites*…"
P12 added that…
"…*modular construction allows for experimentation with new materials and construction techniques*…"

As highlighted by the respondents' comments, one of the main advantages of modular building is its capacity to reduce waste and advance sustainability. Prefabricating building components in a factory setting allows for more effective management and reusing of materials. This process is known as modular construction. This minimises construction waste when compared to conventional on-site building techniques, where supplies are sometimes lost because of bad weather, mistakes made by employees, or overordering. According to the respondents' perspectives, implementing modular construction techniques can reduce costs considerably. Off-site component prefabrication can expedite construction, lower labour costs, and accelerate project completion dates. Additionally, the controlled industrial environment helps construction companies, and their clients save money by minimising material waste and personnel inefficiencies.

8.2.11 Application of Modularity Methodology to Different Building Types

Interviewees were questioned about their opinions on the different types of buildings, and how modularity methodology can be used to develop in the construction industry.

Response:
P1 stated that…
"…*modularity methodology is suitable for all kinds of buildings*…"
P2 opined that…
"…*Residential modularity is tailored for housing projects*…"
P3 remarked that…
"…*Building hotels using modularity methodology makes hospitality services easily accessible*…"
P4 added that…
"…*Mixed-use modularity combines different building functions, such as residential, commercial, and recreational, within a single modular structure to maximise space utilisation and convenience*…"

The modularity methodology, consisting of prefabricated and standardised components installed on-site, is said to apply to various building types by the participants. It implies that a wide range of structures, including residential residences, commercial buildings, and even industrial facilities, can be built using modular construction techniques that can be modified and applied. The modular buildings may effectively meet various practical and architectural needs because of their scalability and flexibility. The interviewees emphasise that modular building techniques can expedite the construction process, shorten the construction schedule, and possibly save money; therefore, they are especially well-suited for housing projects. Residential modularity strongly emphasises the use of prefabricated modules that are readily customisable to match unique housing requirements while upholding standards of quality and longevity.

8.3 Discussion of Interview Results

ISO1 *To identify the main attributes and sub-attributes for modularity methodology in the construction industry and to evaluate whether these attributes would aid the adoption.*

The first objective set for the interview was to identify the main attributes of modularity methodology for the construction industry and to evaluate if the features would aid adoption in the context of emerging economies. The result from the interview study indicates six existing attributes for adopting modularity methodology for the construction industry. The interviewees agreed to all the attributes; hence, there was a consensus on the significance of the attributes. This aligns with other studies on adopting modular construction (Bertram et al., 2019; Gosling et al., 2016; Liew et al., 2019). Modularity methodology, a concept associated with lean production, has become an attractive approach to addressing the usual problems and decelerating the dynamics of the construction industry. This approach complies with seeking organisational transformation and generating knowledge, skills, and innovation that increase competitiveness and scope for obtaining sustainable economic growth and development within the construction industry. These findings show that supply chain, productivity, digitalisation, legal and regulatory considerations, sustainability, and procurement methods are critical determinants in adopting modularity methodology for the construction industry.

For supply chain attributes, twenty-four sub-attributes were generated from the interview. The interviewees mentioned the constant change in the supplier's landscape, support for integrated working styles, conversion of used material to a new product, stakeholders' decisions, and corrective actions when there is deviation. These findings corroborate those of (Abdelmageed & Zayed, 2020; Abdul & El-adaway, 2020; Thai et al., 2020), which reveal that supply chain design may be regarded as one of the crucial concerns in the application of modularity. The distinguishing feature of modularity and the underpinning of its numerous benefits is the capability to form numerous substitutes and critically substitute components into the ultimate assembled product from these formed substitutes. When modularity is applied in the construction industry, supply chain considerations are the primary considerations that affect modularity attributes. It is essential for the decision-wave modularity method to delay which level of supply chain considerations should be used to make the industry-wide decision-level decisions or as a natural result of the component types considered for inclusion or exclusion from product compositions.

Regarding productivity attributes, twenty-one sub-attributes were identified by the interviewees. These include practical risk assessment, project cost, comprehensive safety measures, ample availability of labour, job size and complexity. Yadav and

Kishan (2020) and Ali et al. (2020) established that modularity methodology, which involves buildings in modules or 2D panels, enables the product's cognitive reach to become much more comprehensive with less risk involvement. As constructors have labelled the construction sector for making mistakes in architectural design, coordination, and teamwork execution, several design errors may result in adverse quality and safety effects. However, this scenario can be mitigated by introducing modularity methodology into construction practices (Zhang et al., 2024). This highly reduces errors during construction as technologically advanced precision-based errors are well-identified in sub-workshops or on site to completion, which can raise teamwork execution.

The assessment of the attribute relating to digitalisation had twenty-four sub-attributes. The interviewees mentioned the availability of hardware technology, use of system software, effective data transmission, strategic planning of data management, and effective data governance as the significant sub-attributes. The construction industry is lagging behind in digitalisation and needs radical changes. However, modular construction is the megalith of the construction industry that can change this scenario (Zhou et al., 2021). The manufacturing industry has already introduced modularity and prefabrication to enhance productivity and improve quality. Modularity can introduce a concept of leaner construction and Construction 4.0 such as Industry 4.0. Modularity methodology uses enhanced software such as BIM to optimise costs, time, and resources, as the entire project components are manufactured, verified, and stored digitally before construction starts (Darko et al., 2020; Li et al., 2022a). Structures are manufactured under comfortable conditions, assuming high-level tolerances and avoiding environmental exposure. The designer can opt for innovative framing solutions and incorporate better thermal and separation technologies and innovative supply systems. Robotic machinery combinations also reduce waste from boards and blocks (Liang et al., 2023; Adekunle et al., 2023). The strategy thus minimises peripheral enclosure and the consumption of expensive technical plant systems.

Furthermore, the legal and regulatory factor had fifteen sub-attributes with clear and definite offers, availability of written constitution, preserving individual and organisational rights, unequivocal and corresponding acceptance, and unequivocal and corresponding acceptance mentioned more. This is in line with the findings of Assaad et al. (2020) and Liu et al. (2023) that to avoid future problems with relocating the building units and upgrading them to off-site jurisdictions, the units must satisfy the code and other relevant aspects. The modular perspective structure will be designed to ensure optimum transport and site access while local regulations dictate the need for a crane and construction-related activities. Generally, when project planning involves units of different heights, delivery planning is a significant issue because of the variety of modules (Messaoudi & Nawari, 2020; Razkenari et al., 2020). Land issues include property rights, utilities, and infrastructure. Owners who use off-site design planning lack the comprehensive knowledge that necessary experts must acquire to understand the role of current and future knowledge in the processes exemplified by mini housings. Yang et al. (2021) argued that owing to its decentralised nature, the design team must be familiar with transportation and logistics to provide a means of delivering the unit to the destination. Hussein et al. (2023) also mentioned that to promote construction projects' capacity to adopt modularity methodology, in addition to theoretical knowledge, policymakers should explore construction projects' operational characteristics and design street-level decision support strategies to enhance modularity development in the construction industry.

Sustainability is another attribute with twenty-six sub-attributes. Client satisfaction, digital twin application, energy efficiency and renewable energy, value creation, and natural ventilation were highly mentioned sub-attributes. This is in agreement with the findings of Abdelmageed & Zayed (2020) that the concept of sustainability could be related to design flexibility, ease of disassembly, short modular construction period, high energy efficiency in operation, low on-site waste, robust building components resistant to natural perils, reduced global warming, and low cost. Thai et al. (2020) also proposed a lean operation system in prefabrication to increase its profit and effect on waste. It is pointed out that the lean construction concept has significant relationships with sustainability performance. Hossain et al. (2020) and Wuni & Shen (2020a) indicated that modular construction could lead to design flexibility, zero waste, lower prices, and shorter time, making the designers give primary environmental considerations. Along with the heat dissipation and thermal conduction advantages, the cost, lighting condition, and building weight advantages brought by the steel frame structure configure the modular design to be a creative design within the footprint, elevating the design flexibility and ease of disassembly. Therefore, modular construction has a high potential to develop into a sustainable construction method.

In addition, the procurement method had fourteen sub-attributes with design and build, negotiated, turnkey and more. This agrees with significant studies that stated that choosing the most efficient procurement method is a critical decision that significantly influences the modularity of built environment projects (Wuni & Shen, 2020b; Iacovidou et al., 2021; Akinradewo et al., 2023). Jang et al. (2022) argued that the shift toward modularity methods is a growing trend in the construction industry, as it has excellent potential to enhance innovation and cost savings. While different procurement methods approach projects differently, turnkey negotiated PPPs are found to be decisive in influencing the design process and construction bill of quantities. A reducing strategy is realised through modularity methods, which can be influenced in the design process and construction activities. Projects need to foster a close relationship among participants with clearly defined roles and responsibilities to derive the benefits of modularity methodology.

Another objective was to determine whether one of the gaps identified, namely reassembly, influences modularity methodology for the construction industry. Interviewees suggested that high-quality final products, access and equity, the usage of backup systems and components, meeting clients' needs and expectations, and the ability to maintain structural integrity are the most vital sub-attributes concerning reassembly. This aligns with the findings of Xu et al. (2024) and López Ruiz et al. (2020) that the goal of designing buildings for disassembly is to use reusable products in new construction, minimise waste, promote material recycling, expand the demolition market, and minimise greenhouse gases emitted from the construction industry. However, the obstacles to disassembly must first be overcome.

In contrast, Sharma et al. (2022) argued that the obstacles to disassembly are that building projects do not recognise the value of cost-effective disassembly in advance, and apparent demand can stimulate the building disassembly supply chain. Reassembly is a fundamental strategy in waste management (O'Grady et al., (2021). Products are increasingly designed for disassembly; however, buildings are designed with little consideration for disassembly, making it challenging to manage building waste (Sanchez et al., 2020). Mahmoudi and Nourzard (2022) further describe disassembly as the only waste management option that presents an opportunity for continual growth, provides economic and environmental benefits, and presents disassembly as the only waste management option that minimises waste and greenhouse gases, contributes to sustainable resource management, and minimises environmental degradation.

ISO2 *To determine the influence of reassembly on modularity methodology for the construction industry.*

The second objective was to determine whether one of the gaps identified, namely reassembly, influences modularity methodology for the construction industry. Interviewees suggested that high-quality final products, access and equity, the usage of backup systems and components, meeting clients' needs and expectations, and the ability to maintain structural integrity are the most vital sub-attributes concerning reassembly. This aligns with the findings of Xu et al. (2024) and López Ruiz et al. (2020) that the goal of designing buildings for disassembly is to use reusable products in new construction, minimise waste, promote material recycling, expand the demolition market, and minimise greenhouse gases emitted from the construction industry. However, the obstacles to disassembly must first be overcome.

In contrast, Sharma et al. (2022) argued that the obstacles to disassembly are that building projects do not recognise the value of cost-effective disassembly in advance, and apparent demand can stimulate the building disassembly supply chain. Reassembly is a fundamental strategy in waste management (O'Grady et al., (2021). Products are increasingly designed for disassembly; however, buildings are designed with little consideration for disassembly, making it challenging to manage building waste (Sanchez et al., 2020). Mahmoudi and Nourzard (2022) further describe disassembly as the only waste management option that presents an opportunity for continual growth, provides economic and environmental benefits, and presents disassembly as the only waste management option that minimises waste and greenhouse gases, contributes to sustainable resource management, and minimises environmental degradation.

ISO3 *To determine the influence of maintenance on modularity methodology for the construction industry.*

The third objective was to determine whether one of the identified gaps, namely maintenance, influences modularity methodology for the construction industry. Interviewees suggested characterisation of the risk, minimising overtime, identifying hazards, and minimising accidents and minimising breakdown as the most vital sub-attributes concerning maintenance. This agrees with the opinion that the maintenance of the property after its delivery by the building service is a general concern since there must be adequate conditions for efficient management, as observed by Zheng et al. (2024). The choice of the materials and techniques used in the construction process must be as efficient as possible. Abdul Nabi and El-adaway (2020) and Iacovidou et al. (2021) argued that maintenance is required for residential properties and should be performed to prevent worsening conditions due to the wear and tear of materials, which can lead to high costs, generating great dissatisfaction on the part of users. Ideally, in the pre-project, a choice must be made for the materials and constructive solutions that have a long service life and the ones that promote comfort in the internal environment with a focus on the quality of housing (Abdelmageed & Zayed, 2020; Loizou et al., 2021). Full compliance after delivery by construction professionals requires commitment and a proposal for new solutions, making the construction sector rethink its concepts, techniques, and design methodologies, incorporating industrial production processes in the face of increasingly demanding heterogeneities. Modularity methodology is an alternative for change in the productive system (Olawumi et al., 2022). Therefore, it seeks to provide and present solutions for developing companies and services in the construction sector. However, resistance to methodological change poses a significant barrier to progress in this field. Moreover, maintenance is an essential factor in the core business of the construction of buildings or properties.

ISO4 *To determine the influence of cultural perception on modularity methodology for the construction industry.*

The fourth objective was to determine whether one of the identified gaps, namely cultural perception, influences modularity methodology for the construction industry. Interviewees suggested that resource management practices, understanding construction methods through sighting, Indigenous construction knowledge, practical, functional, and efficient construction approach, and knowledge-based construction decision-making and execution are the most vital sub-attributes concerning cultural perception. These results are consistent with those of Iacovidou et al. (2021) and Loizou et al. (2021), who found that cultural beliefs might influence attitudes towards innovation and change in the construction industry. Adopting modularity methodology may encounter resistance in regions where traditional construction methods are well ingrained. The popularity of the modularity approach in the construction sector can be influenced by cultural variables, such as a preference for familiar procedures or risk aversion. Xu et al. (2020) contended that although hierarchical structures and collaborative techniques may coexist in various organisations, the implementation of modular methodologies may differ based on the customary decision-making processes in each setting. Modularity approaches must be flexible enough to accommodate local circumstances, such as societal norms, environmental factors, and legal constraints (Thai et al., 2020; Wuni & Shen, 2020a). Modular projects can be more acceptable and efficient by understanding and incorporating these components into the planning and implementation process. Construction methods can be encouraged as time-saving mechanisms by effectively using limited resources. Field tasks now come from financial means transferred to over-educated men and women. Increased financial collaboration for non-virtual work movements also provides more opportunities for non-virtual workforce movements.

ISO5 *To determine the extent to which modularity methodology can be adopted for the different types of buildings.*

The fifth objective of the interview was to determine how modularity methodology can be adopted for the different types of buildings in South Africa. The different types of buildings have their intended functions and uses depending on the set-out plans and ideas for which they are initially constructed. This ranges from residential buildings for housing its populace, industrial facilities and layouts for producing goods to healthcare infrastructure that caters for healthcare delivery, and storage facilities that aim to house and store unfinished and finished goods for short- or long-term purposes. Other facilities are religious facilities that aid in gathering people for religious activities; hospitality facilities cover infrastructure from hotels to spas. It is imperative to note that the espousal and utilisation of modularity methodology in different buildings will be driven in varying dimensions. Since the various types of buildings all have various functions, modularity methodology might be of good use in optimising process delivery in some types of facilities compared to others. Adopting modularity methodology might bring comparative advantages in some facilities, such as monitoring physical assets, managing workflow processes, overseeing overall building facilities, and maintaining maintenance. Some types of buildings would be better suited to incorporate the methodology than others. Given those, as mentioned earlier, it was essential to determine the extent to which modularity methodology can be adopted to construct the different types of buildings.

Findings from the study showed that among the thirteen types of buildings identified, the buildings that proved to be best suited for modularity methodology are residential, office, industrial, healthcare, hospitality, and storage. The reason for such opinions from the experts is not farfetched. Since modularity methodology is one of the conveyors of the 4IR, anchored on overhauling production and manufacturing processes, it is imperative to establish that using such methods in this type of building is not out of place. Also, the improvement of value proposition exhibited by the traditional industry, coupled with the creation of opportunities and business models that are entirely new. Data analysis for improving performance is one of the comparative advantages of inculcating an innovative building practice such as modularity methodology. It was also found that the modularity concept will mainly benefit healthcare buildings. Healthcare facilities frequently have to adjust to the shifting requirements of their patients and changes in healthcare delivery methods and technological improvements. Because of its modular design, spaces can be more easily enlarged or altered to meet changing needs without significantly disrupting ongoing business operations.

ISO6 *To predict the extent to which the various outcomes of modularity methodology can serve as drivers for adopting the construction method.*

The last objective of the interview was to predict the extent to which the various outcomes of modularity methodology can serve as drivers for adopting the construction method in South Africa. It has been noted that modular construction involves a broad spectrum of functions, activities, and operations, often referred to as the scope of construction management. However, it can be attested that some of these outcomes would drive the implementation of modularity methodology more than others. Among the seventeen identified modularity methodology outcomes, it was shown in the study that eleven of the outcomes are to a great extent in the drive for modularity methodology, while six outcomes are to a great extent. As revealed in the interview, 'Efficiency and quality control', 'Reduction in material waste', 'Reduction in life-cycle embodied energy and carbon', 'Reduction in disruption to the surrounding community,' and 'Speedy construction' are the functions that serve as very high drivers for the espousal of modularity methodology for the construction industry. This is corroborated by Chauhan et al. (2022) and Hackl et al. (2020), who stated that applying the modularity methodology in the construction process can lead to cost-efficient benefits.

For higher efficiency and cost-effective solutions, design engineers will refer to familiar aspects of individual requirements rooted in a product lifecycle to assemble a standard product. Zhao and Tan (2024) opined that as several customers' demand patterns are considered and understood, contractors can optimise standard options and subsequently create better elements and processes that considerably reduce the time and costs associated with these processes. The product's quality is open to continual improvement, and the components can be modified, updated, removed, or replaced depending on how the demand from the market evolves. Moreover, when the construction process is undertaken under strict factory control or in a supervised form, which is more stringent than traditional construction, it gives certainty about the performance standards (Meng et al., 2021; Darko et al., 2020). Applying the modularity methodology in in-ground and out-ground slab construction has proven to be a cost-effective method that can reduce construction time, produce standard construction quality, and reduce environmental emissions. The modularity methodology in construction involves transforming and replicating the construction process to construct a standard product (Adekunle et al., 2024). It is produced by assembling the prefabricated standardised individual components internally, vertically, and horizontally.

In addition, Loizou et al. (2021) stated that with the current challenge of the disposal of construction waste, modularity methodology in construction is considered one of the possibilities for reducing material waste. Combining timber as a construction material, modular construction allows detailed management within a factory-controlled environment, such as off-site construction (Iacoividou et al., 2021). Many buildings designed and built in the past have been demolished a few decades later. Such demolition waste often contains hazardous elements and heavy metals; therefore, it is recommended that it be minimised. It is crucial to broaden our design approach to make buildings more valuable. Pervez et al. (2022) revealed that the reuse of construction materials is required, so effective measures should be taken to preserve materials in the future. Such as reducing the use of composite construction products with hazardous elements and the amount of materials used to increase the reuse of construction materials; at the same time, the increase in the volume of waste created will lly lead to an increase in future land displacement operations. By valuing, designing, and managing sustainable reuse and mineralisation, modular construction can continue to create reduced environmental impacts, savings in both material and financial terms, and other benefits while overcoming the negative perception of specific users (Abdul Nabi & El-Adawy, 2020; Nam et al., 2020).

8.4 Implication of Findings

A notable implication from the study's findings is that for modularity methodology in the construction industry to be successfully and adequately carried out; significant emphasis should be placed on the system's procurement method. These are private financing, competitive bidding, negotiated contracts and reverse auction. Davis et al. (1989) and Moore (2013) corroborated this view by stating that it would be essential to inculcate an effective procurement strategy in modular project delivery when there is no guarantee of making significant improvements with the procurement system being deployed. In this light, it is safe to state that the assurance of improved construction procedures and deliveries is critical to adopting modern technological methodology, such as modularity methodology. Since it has been proven that modularity methodology guarantees significant competitive advantage due to its attributed efficiencies and streamlined processes, this approach is highly recommended. Beyond the expected benefits of deploying modularity methodology in construction, the focus should also be on digitalisation as a significant determinant for its adoption. The attributes of digitalisation include deriving value from data assets, effective data governance, organisational culture relating to data management, access and sharing of data, quality and integrity of data, and effective integration of changes.

Furthermore, the supply chain attributes needed for using a construction method such as modularity methodology also prove to be a very vital attribute for its adoption. Issues such as forecasting unexpected twists (Pagell & Wu, (2009), accurate view of order history (Behera et al., 2015), stakeholder's decision (Gosling et al., 2015), converting used material to a new product (Li et al., 2022b), use of road mapping tools (Quang et al., (2017), pursuing right initiatives (Studer & De Brito Mello, 2021), reallocating a new function to materials (Zaalouk et al., 2023) and converting used material to a new product (Pero et al., 2010) should receive high levels of consideration. Forecasting unexpected twists involves predicting and preparing for unforeseen challenges during construction. Modular construction could include delays in module delivery, changes in design specifications, or disruptions in the supply of materials. Accurate forecasting helps create contingency plans, maintain project timelines, and ensure that resources are available when needed. According to Badawy et al., (2025), predictive modelling and sophisticated analytics tools can foresee possible problems and create mitigation plans. Planning future orders, keeping an accurate perspective of order history, and guaranteeing that materials are available when needed all depend on inventory management.

When the building industry adopts the modularity methodology, which involves prefabricated off-site components, knowing historical order trends might assist in streamlining procurement procedures and cutting waste. With the help of this historical data, supply chain managers can make smarter decisions and operate their businesses more effectively by learning more about lead times, material usage rates, and supplier performance. Additionally, Wuni and Shen (2020b) contended that stakeholder decisions significantly influence the adoption of modularity methodology. Clients, architects, engineers, contractors, and suppliers are examples of stakeholders who can affect the results of a project. According to O'Connor et al. (2014), collaborative decision-making guarantees that all stakeholders are committed to the success of the modular approach, understand its advantages, and are aligned with it. Early stakeholder engagement and open lines of communication are vital to resolving issues, winning support, and promoting more efficient project management. Converting spent material into new goods requires recycling and reusing materials from earlier projects, which is essential for smoother project execution. This approach is consistent with modular construction concepts and promotes sustainability and waste reduction. Reusing elements such as wooden panels, steel frames, and other materials makes it possible to create new modules. Recycling processes and technologies can be implemented more efficiently by ensuring that materials are reused effectively and lowering the costs and environmental impact of purchasing new materials.

According to Abdul Nabi and El-adaway (2020), road mapping technologies are crucial for organising and supervising the modular construction process. These tools visually represent the project schedule, critical checkpoints, and task dependencies. They support the coordination of all project phases, resource alignment, and the identification of possible bottlenecks. Road mapping tools also help stakeholders communicate with one another by giving them a clear picture of the project's status and any modifications required to keep it on course. These tools make it easy for all stakeholders to comprehend and align with the project's objectives by visualising the project timeline, important milestones, and task interdependencies.

8.5 Contextualising the Validated Modularity Methodology

The modularity adoption model for the construction industry was validated, indicating the relationship between the exogenous variables and the endogenous variable. Based on the preceding contextualisation of the validated model, for a well-encompassing modularity methodology to be attained for the functions and activities of the construction industry, the general attributes for the adoption are grouped under two major headings, namely intrinsic and extrinsic features as seen in Fig. 8.1. The intrinsic features refer to inherent modalities required for the system's adoption.

These are the congenital and innate attributes to be considered for driving modularity methodology for the construction industry. This category has six sub-attributes: supply chain, productivity, digitalisation, legal and regulatory consideration, sustainability, and procurement methods. As noted from the extant literature review and the results of the interview carried out. These attributes result from the envisaged outcomes of the use of the method, perceptions or beliefs of societal stances resulting from its use, the enabling conditions required for its use, and the necessary working attributes for the effective use of the system. Hence, it can be said that these attributes can be classified as intrinsic factors serving as measures for adopting the construction method.

The other main attribute is categorised as extrinsic features. This refers to the external modalities needed or required for adopting the construction method. These attributes can be attributed to extraneous or external characteristics that can serve as propelling measures for adopting modularity methodology for the construction industry. Under this category, there are three sub-attributes: reassembly, maintenance

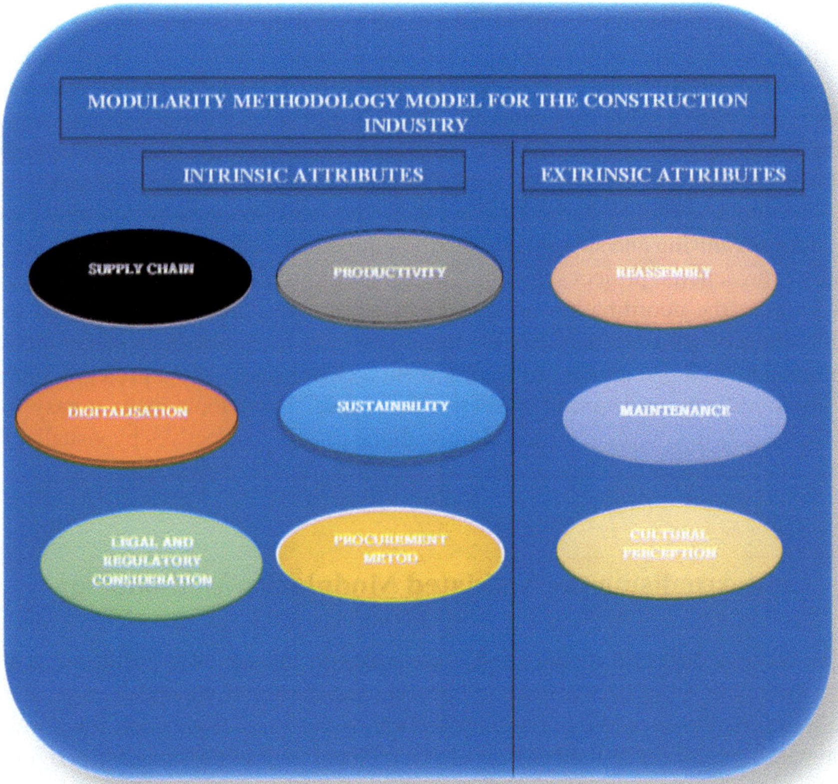

Fig. 8.1 Modularity methodology model for the construction industry

and cultural perception. Also, as observed from the review of extant literature and the results from the interview, these attributes result from external factors serving as drivers for the methodology adoption. They dwell on the strategies put in place by organisations for effective and efficient operations and deliveries based on economic factors and the quantification of the deliveries and output made by the method. Therefore, these attributes can be ascribed to being extrinsic.

Furthermore, adopting a modularity methodology, affected by these three main attributes when properly integrated based on the validated model, would yield the much-desired outcomes from the method's adoption. These are efficiency and quality control, reduction in life-cycle embodied energy and carbon, reduction in energy use for construction, support of adaptation, reuse and recycling, support of resilience, reduction in disruption to the surrounding community, minimal negative impact on business, speedy construction and flexibility of use.

8.6 Summary

The presentation of the outcomes of the interviews conducted is made in this chapter. The analyses were carried out based on answers to the individual questions posed to the panel of experts. These questions dwelt on the significance of the main and sub-attributes for adopting modularity methodology for the construction industry. Also, the extent to which modularity methodology can be adopted for the different types of buildings was determined. Likewise, the prediction of the extent to which the various outcomes of modular construction can serve as drivers for modularity methodology was presented. A comprehensive discussion of the study's findings concerning the outlined objectives of the study concluded the chapter. The highlights of the findings showed coherency regarding the experts' consensus concerning modularity methodology for the construction industry. Ultimately, the findings of the interview project the key determinants and constructs in the determination of modularity methodology for the construction industry.

References

Abdelmageed, S., & Zayed, T. (2020). A study of literature in modular integrated construction-Critical review and future directions. *Journal of Cleaner Production, 277*, 124044. https://doi.org/10.1016/j.jclepro.2020.124044

Abdul Nabi, M., & El-adaway, I. H. (2020). Modular construction: Determining decision-making factors and future research needs. *Journal of Management in Engineering, 36*(6). https://doi.org/10.1061/(asce)me.1943-5479.0000859

Adekunle, P., Aigbavboa, C., Otasowie, K. & Adekunle, S. (2023). Benefits of robotic utilization in the prefabricated construction industry. In *Proceedings of the 31st Annual Conference of the International Group for Lean Construction (IGLC31)* (pp. 746–754). https://doi.org/10.24928/2023/0134

Adekunle, P., Aigbavboa, C., Otasowie, K., & Akinradewo, O. (2024). Matching-up modularity methodology application within the built environment: A bibliometric review. *Journal of Asian Architecture and Building Engineering*, 1–16. https://doi.org/10.1080/13467581.2024.2368653

Akinradewo, O., Aigbavboa, C., Aghimien, D., Oke, A., & Ogunbayo, B. (2023). Modular method of construction in developing countries: The underlying challenges. *International Journal of Construction Management, 23*(8), 1344–1354. https://doi.org/10.1080/15623599.2021.1970300

Assaad, J. J., Abou Yassin, A., Alsakka, F., & Hamzeh, F. (2020). A Modular approach for steel reinforcing of 3D printed concrete—Preliminary study. *Sustainability, 12*(10), 4062. https://doi.org/10.3390/su12104062

Badawy, M., Ibrahim, A., Hamdy, K., & Aljohani, A. (2025). A comparative analysis of barriers to prefabricated construction adoption in emerging economies: a dual fuzzy-CB-SEM and PLS-SEM approach. *International Journal of Construction Management*, 1–21. https://doi.org/10.1080/15623599.2025.2569647

Behera, P., Mohanty, R. P., & Prakash, A. (2015). Understanding construction supply chain management. *Production Planning & Control, 26*(16), 1332–1350. https://doi.org/10.1080/09537287.2015.1045953

Bertram, N., Fuchs, S., Mischke, J., Palter, R., Strube, G., & Woetzel, J. (2019). Modular construction: From projects to products. *Capital Projects & Infrastructure*, 1–30. https://www.mckinsey.com/industries/capital-projects-and-infrastructure/our-insights/modular-construction-from-projects-to-products

Chauhan, K., Peltokorpi, A., Lavikka, R., Seppänen, O., Chauhan, K., Peltokorpi, A., Lavikka, R., & Seppänen, O. (2022). The monetary and non-monetary impacts of prefabrication on construction: The effects of product modularity. *Buildings, 12*(4). https://doi.org/10.3390/buildings12040459

Darko, A., Chan, A. P., Yang, Y., & Tetteh, M. O. (2020). Building information modeling (BIM)-based modular integrated construction risk management—Critical survey and future needs. *Computers in Industry, 123*, 103327. https://doi.org/10.1016/j.compind.2020.103327

Davis, K., Ledbetter, W. B., & Burati Jr, J. L. (1989). Measuring design and construction quality costs. *Journal of Construction Engineering and Management, 115*(3), 385-400.

Gosling, J., Pero, M., Schoenwitz, M., Towill, D., & Cigolini, R. (2016). Defining and categorizing modules in building projects: An international perspective. *Journal of Construction Engineering and Management, 142*(11), 1–11. https://doi.org/10.1061/(asce)co.1943-7862.0001181

Gosling, J., Towill, D. R., Naim, M. M., & Dainty, A. R. J. (2015). Principles for the design and operation of engineer-to-order supply chains in the construction sector. *Production Planning & Control, 26*(3), 203–218. https://doi.org/10.1080/09537287.2014.880816

Hackl, J., Krause, D., Otto, K., Windheim, M., Moon, S. K., Bursac, N., et al. (2020). Impact of modularity decisions on a firm's economic objectives. *Journal of Mechanical Design, 142*(4), 041403.

Hossain, M. U., Ng, S. T., Antwi-Afari, P., & Amor, B. (2020). Circular economy and the construction industry: Existing trends, challenges and prospective framework for sustainable construction. *Renewable and Sustainable Energy Reviews, 130*, 109948. https://doi.org/10.1016/j.rser.2020.109948

Hussein, M., Karam, A., Eltoukhy, A. E.E., Darko, A., & Zayed, T. (2023). Optimized multi-modal logistics planning of modular integrated construction using hybrid multi-agent and metamodeling. *Automation in Construction 145*, 104637. https://doi.org/10.1016/j.autcon.2022.104637

Iacovidou, E., Purnell, P., Tsavdaridis, K. D., & Poologanathan, K. (2021). Digitally enabled modular construction for promoting modular components reuse: A UK view. *Journal of Building Engineering, 42*, 102820. https://doi.org/10.1016/j.jobe.2021.102820

Jang, H., Ahn, Y., Roh, S., Jang, H., Ahn, Y., & Roh, S. (2022). Comparison of the embodied carbon emissions and direct construction costs for modular and conventional residential buildings in South Korea. *Buildings, 12*(1). https://doi.org/10.3390/buildings12010051

Li, C. Z., Li, S., Li, X., Wu, H., Xiao, B., Y. Tam, V. W., et al. (2022a). A scientometric review of management of prefabricated construction from 2011–2021. *Buildings, 12*(10). https://doi.org/10.3390/buildings12101515

Li, X., Lu, W., Xue, F., Wu, L., Zhao, R., Lou, J., et al. (2022b). Blockchain-enabled IoT-BIM platform for supply chain management in modular construction. *Journal of Construction Engineering and Management, 148*(2), 04021195.

Liang, R., Li, R., & Chong, H. Y. (2023). Sustainable supplier selection for prefabricated megaprojects during COVID-19: A novel large group decision-making model. *Building Research & Information, 51*(5), 533–554. https://doi.org/10.1080/09613218.2023.2165474

Liew, J. Y. R., Chua, Y. S., & Dai, Z. (2019). Steel concrete composite systems for modular construction of high-rise buildings. *Structures, 21*, 135–149. https://doi.org/10.1016/j.istruc.2019.02.010

Liu, C., Song, Y., Li, R., Ma, W., Hao, J. L., & Qiang, G. (2023). Three-level modular grid system for sustainable construction of industrialized residential buildings: A case study in China. *Journal of Cleaner Production, 395*, 136379. https://doi.org/10.1016/j.jclepro.2023.136379

Loizou, L., Barati, K., Shen, X., Li, B., Loizou, L., Barati, K., et al. (2021). Quantifying advantages of modular construction: Waste generation. *Buildings, 11*(12). https://doi.org/10.3390/buildings11120622

López Ruiz, L. A., Roca Ramón, X., & Gassó Domingo, S. (2020). The circular economy in the construction and demolition waste sector—A review and an integrative model approach. *Journal of Cleaner Production, 248*, 119238. https://doi.org/10.1016/j.jclepro.2019.119238

Mahmoudi Motahar, M., & Hosseini Nourzad, S. H. (2022). A hybrid method for optimizing selective disassembly sequence planning in adaptive reuse of buildings. Engineering, *Construction and Architectural Management, 29*(1), 307–332. https://doi.org/10.1108/ECAM-12-2020-1023

Meng, Q., Liu, W., Li, Z., & Hu, X. (2021). Influencing factors, mechanism and prevention of construction workers' unsafe behaviors: A systematic literature review. *International Journal of Environmental Research and Public Health, 18*(5), 2644. https://doi.org/10.3390/ijerph18052644

Messaoudi, M., & Nawari, N. O. (2020). Virtual permitting framework for off-site construction case study: A case study of the state of Florida. *Proceedings of the 18th International Conference on Computing in Civil and Building Engineering: ICCCBE 2020* (pp. 757–771).

Moore, D. R. (2013). Construction. In *Innovations for sustainable building design and refurbishment in Scotland: The outputs of CIC start online project* (pp. 87–104). Springer International Publishing.

Nam, S., Yoon, J., Kim, K., & Choi, B. (2020). Optimization of prefabricated components in housing modular construction. *Sustainability, 12*(24), 10269. https://doi.org/10.3390/su122410269

O'Grady, T. M., Brajkovich, N., Minunno, R., Chong, H.-Y., & Morrison, G. M. (2021). Circular rconomy and virtual reality in advanced BIM-based prefabricated construction. *Energies, 14*(13), 4065. https://doi.org/10.3390/en14134065

Olawumi, T. O., Chan, D. W. M., Ojo, S., & Yam, M. C. H. (2022). Automating the modular construction process: A review of digital technologies and future directions with blockchain technology. *Journal of Building Engineering, 46*, 103720. https://doi.org/10.1016/j.jobe.2021.103720

Pagell, M., & Wu, Z. (2009). Building a more complete theory of sustainable supply chain management using case studies of 10 exemplars. *Journal of supply chain management, 45*(2), 37-56. https://doi.org/10.1111/j.1745-493X.2009.03162.x

Pero, M., Abdelkafi, N., Sianesi, A., & Blecker, T. (2010). A framework for the alignment of new product development and supply chains. *Supply Chain Management: An International Journal, 15*(2), 115–128. https://doi.org/10.1108/13598541011028723

Pervez, H., Ali, Y., Pamucar, D., Garai-Fodor, M., & Csiszárik-Kocsir, Á. (2022). Evaluation of critical risk factors in the implementation of modular construction. *Plos One, 17*, 1–27. https://doi.org/10.1371/journal.pone.0272448

Quang, H. T., & de Castro, R. (2017). Impact of supply chain alignment on construction performance: A developed model for Vietnam. *International Journal of Construction Supply Chain Management, 7*(2), 68–92.

Razkenari, M., Fenner, A., Shojaei, A., Hakim, H., & Kibert, C. (2020). Perceptions of offsite construction in the United States: An investigation of current practices. *Journal of Building Engineering, 29*, 101138. https://doi.org/10.1016/j.jobe.2019.101138

Sanchez, B., Rausch, C., Haas, C., & Saari, R. (2020). A selective disassembly multi-objective optimization approach for adaptive reuse of building components. *Resources, Conservation and Recycling, 154*, 104605. https://doi.org/10.1016/j.resconrec.2019.104605

Studer, W. P., & De Brito Mello, L. C. B. (2021). Core elements underlying supply chain management in the construction industry: A systematic literature review. *Buildings, 11*(12), 569. https://doi.org/10.3390/buildings11120569

Thai, H., Ngo, T., & Uy, B. (2020). A review on modular construction for high-rise buildings. *28*, 1265–1290.

Wuni, I. Y., & Shen, G. Q. (2020a). Barriers to the adoption of modular integrated construction: Systematic review and meta-analysis, integrated conceptual framework, and strategies. *Journal of Cleaner Production, 249*, 119347. https://doi.org/10.1016/j.jclepro.2019.119347

Wuni, I. Y., & Shen, G. Q. (2020b). Critical success factors for modular integrated construction projects: a review. *Building Research and Information, 48*(7), 763–784. https://doi.org/10.1080/09613218.2019.1669009

Xu, Y., Liu, S., & Heisel, F. (2024). Towards sustainable construction waste management: Study on a disassemblable brick partition wall for the architecture, construction, and engineering industry. *Circular Economy, 3*(1), 100078. https://doi.org/10.1016/j.cec.2024.100078

Yadav, S., & Kishan, B. (2020). Assessments of computational intelligence techniques for predicting reliability of componentbased software parameter and design issues. *International Journal of Advanced Research in Engineering and Technology, 11*(6).

Yang, Y., Pan, M., Pan, W., & Zhang, Z. (2021). Sources of uncertainties in offsite logistics of modular construction for high-rise building projects. *Journal of Management in Engineering, 37*(3), 04021011.

Zaalouk, A., Altaf, M. S., & Han, S. (2023). Simulation-based approach for master planning and scheduling in offsite construction supply chain management. *Computing in Civil Engineering, 2023*, 138–145.

Zhang, Y., Shen, G. Q., & Xue, J. (2024). A bibliometric analysis of supply chain management within modular integrated construction in complex project management. *Buildings, 14*(6). https://doi.org/10.3390/buildings14061667

Zhao, Y., & Tan, T. (2024). Sustainable design for manufacture and sssembly: Exploring modularity and material innovation in earth block construction. In *Proceedings of the 28th International Symposium on Advancement of Construction Management and Real Estate.* Springer.

Zheng, Z., Li, Y., & Torres, J. (2024). An integrated method of automated layout design and optimization for modular construction. *Engineering, Construction and Architectural Management, 30*(3), 1016–1036. https://doi.org/10.1108/ECAM-04-2022-0329

Zhou, J. X., Shen, G. Q., Yoon, S. H., & Jin, X. (2021). Customization of on-site assembly services by integrating the internet of things and BIM technologies in modular integrated construction. *Automation in Construction, 126*, 103663. https://doi.org/10.1016/j.autcon.2021.103663

Part IV
General Summary

Chapter 9
Conclusion and Recommendation

Abstract This chapter presents the conclusion and recommendations of the study on the adoption of modularity methodology within the construction industry. The study set out to develop a conceptual adoption model by reviewing existing theories and frameworks, identifying critical attributes and sub-attributes, and validating these through expert interviews. The findings revealed that modularity methodology holds significant potential to enhance construction efficiency, sustainability, and adaptability. However, issues such as cultural perceptions, regulatory gaps, maintenance practices, and procurement strategies remain critical factors in enhancing adoption. The conclusion synthesises the key findings, demonstrating that successful implementation of modular construction requires a holistic framework that integrates technological readiness, stakeholder collaboration, regulatory alignment, and long-term maintenance strategies. Based on these insights, recommendations are proposed to guide policymakers, industry stakeholders, and practitioners in promoting modularity methodology. These include strengthening regulatory frameworks, fostering strategic partnerships, encouraging innovation through research and training, and creating incentives for sustainable practices. The chapter underscores the relevance of modularity methodology in addressing pressing construction industry challenges while providing a roadmap for its effective adoption. The recommendations serve as practical and strategic guidelines for accelerating modular construction in South Africa and beyond.

9.1 Conclusion

9.1.1 Research Objective 1

The study's first objective was to evaluate the factors determining the adoption of modularity methodology in the construction industry based on a literature review. In order to achieve this objective, an extensive review of the literature was conducted. The findings from this review indicate that the adoption of modularity methodology in construction is influenced by a multi-faceted array of attributes, encompassing supply

P. Adekunle et al., *Modularity Methodology*,
https://doi.org/10.1007/978-3-032-12918-5_9

chain, productivity, digitalisation, legal and regulatory considerations, sustainability and procurement methods. Each attribute plays a crucial role in shaping the decision to implement modular construction practices, highlighting the complexity and interdependence inherent in this approach. Further findings from the literature review revealed that modularity methodology is a critical topic within the construction industry and a significant area of focus across various other industries such as manufacturing, computing, automotive, and energy. In manufacturing, modularity aids in streamlining production processes and enhancing product customisation. In the computing industry, it supports the development of scalable and flexible software architectures. The automotive industry benefits from modularity through efficient vehicle design and assembly processes, while the energy sector leverages modularity to facilitate deploying renewable energy solutions and infrastructure projects.

The cross-industry relevance of modularity methodology underscores its versatility and the broad range of applications that can drive innovation, efficiency, and sustainability. Modularity methodology can revolutionise traditional building practices in the construction industry by promoting prefabricated components, reducing construction time, and minimising waste. The literature suggests that successful adoption hinges on integrating advanced technologies, such as digital modelling and intelligent manufacturing, which can enhance precision and coordination. Additionally, regulatory frameworks and industry standards must evolve to support modular construction practices, ensuring safety, quality, and compliance. Procurement strategies within construction organisations and the broader industry are also necessary to embrace modularity fully, fostering an environment that values innovation and adaptability. The literature analysis underscores the significance of a comprehensive strategy towards adopting modularity methodology in the construction industry. This strategy should consider the interplay of several elements and incorporate insights from other sectors that have effectively employed modular methods.

9.1.2 Research Objective 2

The study's second objective was to investigate and integrate the existing theories and literature published on modularity methodology and identify the gaps that need consideration. Investigate and integrate the existing theories and literature published on modularity methodology and identify the gaps that need consideration. A review of literature was carried out to achieve this objective. The findings revealed that modularity methodology in construction research has not been studied with an all-inclusive construct in developing the previous models and theories. The gaps identified from the extensive literature review were reassembly, maintenance, and cultural perceptions. The gaps formed the new construct in the current study conceptual framework. Since these attributes are necessary to improve the sustainability and lifecycle management of modular structures, the gaps that have been observed were deemed critical. These attributes align with the general objectives of decreasing waste, optimising resource efficiency, and encouraging a circular economy within the building sector.

The study clarified how important it is to have a thorough maintenance plan to guarantee the durability and functionality of modular buildings. This means training maintenance personnel on the unique requirements of modular buildings and developing maintenance protocols specific to modular components. The study found that prior models frequently disregarded the influence of cultural perspectives on innovation, standardisation, and sustainability in the building industry. By incorporating cultural perspectives into the conceptual framework, the study attempted to comprehend and handle any opposition to modular building from various stakeholders, including architects, contractors, clients, and regulatory agencies. Therefore, the current study offers a synthesised classification of the constructs, which should be collectively considered to predict the adoption of modularity methodology in the construction industry. From the synthesised literature, this current study argues that modularity methodology for the construction industry is a nine-factor construct.

9.1.3 Research Objective 3

Findings further revealed that modularity methodology can be used to construct various types of buildings. This approach is well-suited to addressing the unique challenges and opportunities within the construction industry. Modularity methodology offers significant advantages for residential construction, including the ability to produce high-quality, affordable housing quickly and efficiently. In South Africa, where there is a pressing need to address housing shortages, modular construction can accelerate the delivery of residential units, especially in urban and peri-urban areas. Prefabricated modules can be manufactured off-site, transported, and assembled on-site, reducing construction time and costs while maintaining high-quality and durability standards.

Modularity methodology provides flexibility and scalability for commercial buildings, such as office complexes, retail centres, and hotels. Businesses can benefit from shorter construction times, allowing for quicker occupancy and revenue generation. Additionally, the modular approach facilitates future expansions or modifications, as modules can be added or reconfigured to meet changing needs. This adaptability is particularly beneficial in dynamic urban environments where space and demand fluctuate. Modular construction is also well-suited for educational facilities, including schools, universities, and training centres. The ability to construct classrooms, dormitories, and administrative buildings rapidly can help address the growing demand for educational infrastructure. Modular buildings can be designed to meet specific educational requirements, ensuring conducive learning environments. Furthermore, the portability of modular units allows for the establishment of temporary or semi-permanent educational facilities in remote or underserved areas.

9.1.4 Research Objective 4

The study's fourth research objective was to develop a modularity methodology adoption model for the construction industry based on literature. An extensive literature review was used to achieve this objective. The conceptual model theorised that modularity methodology adoption for the construction industry is a nine-factor construct. These factors were supply chain, productivity, digitalisation, legal and regulatory considerations, sustainability, procurement method, reassembly, maintenance and cultural perception. This hypothesis was validated through an interview survey with experts on modular construction.

9.1.5 Research Objective 5

The fifth objective of the study was to determine the validity of the model of modularity methodology for the construction industry. The validation was carried out on the study's conceptualised modularity methodology model for the construction industry. The development of the model was based on a review of extant literature, which helped establish the model's exogenous and endogenous variables. Findings from the literature review indicated that the main attributes for adopting modularity methodology for the construction industry are supply chain, productivity, digitalisation, legal and regulatory considerations, sustainability, procurement method, reassembly, maintenance and cultural perception. The extent of influence and the validity of the hypothesised model were established by experts' opinions derived from the interviews used for the research. Findings from the analysis showed that the nine exogenous variables all had a significant relationship with the outcomes of adopting modularity methodology for the construction industry. At the same time, the procurement method proved to be the most significant relationship with the outcome variables. Furthermore, the study categorised the adoption attributes into two groups. The first group, the intrinsic attributes for adoption, had six attributes: supply chain, productivity, digitalisation, legal and regulatory considerations, sustainability and procurement methods. The second group, which comprises extrinsic attributes, had three attributes: reassembly, maintenance, and cultural perception.

9.2 Recommendation

The following recommendations are made based on the findings of the study:

- Comprehensive training programmes should be developed to address the skills gap and ensure that personnel are well versed in modular construction techniques. These programmes should cover the principles of modularity, specific construction techniques, and the use of relevant technologies such as BIM.

- Improving supply chain management is critical for the successful modularity methodology. Firms should establish strong relationships with suppliers and manufacturers and implement robust forecasting and inventory management systems to ensure the timely delivery of modular components.
- Adopting advanced digital tools and technologies can streamline the modular construction process. Tools such as BIM, project management software, and digital twins should be integrated to enhance project planning, coordination, and real-time monitoring.
- Standardising modular components and ensuring interoperability across different systems and platforms can significantly reduce costs and complexity. Industry-wide standards should be developed and adopted to facilitate the seamless integration of modular components from various suppliers.
- Sustainability should be a core focus of modular construction practices. Firms should prioritise using eco-friendly materials, energy-efficient designs, and waste-reduction strategies. Additionally, the reassembly and recycling of modular components should be promoted by government establishments to support the circular economy.
- To support the widespread of modularity methodology, it is essential to update and strengthen legal and regulatory frameworks. Policies and regulations should encourage innovation, provide clear guidelines for modular construction, and address any potential barriers to its implementation.
- Cultural perceptions play a significant role in the adoption of new methodologies. Efforts should be made to educate stakeholders about the benefits of modular construction, address misconceptions and demonstrate its advantages through pilot projects and success stories. Encouraging a cultural shift towards acceptance of modularity is crucial.
- Incorporating reassembly and maintenance considerations into modular projects' design and planning stages can enhance their long-term viability. Maintenance protocols should be established, and modular components should be designed for easy disassembly and reuse, ensuring sustainability and cost-effectiveness.
- Procurement methods should be adapted to support modular construction. Strategies such as reverse auction procurement can be employed to ensure competitive pricing and high-quality components. Contracts should clearly define the roles and responsibilities of all parties involved, with a focus on collaboration and efficiency.
- Effective project management and coordination are essential for modular construction projects. Firms should adopt integrated project delivery methods and foster strong collaboration among architects, engineers, contractors, and suppliers to ensure smooth project execution.
- Governments can play a crucial role in promoting modularity methodology. Supportive policies, financial incentives, and funding for research and development can encourage firms to invest in modular construction practices.

- Continuous research and development are needed to advance modular construction technologies and methodologies. Firms and academic institutions should collaborate to explore innovative solutions, address existing challenges, and improve the overall efficiency and effectiveness of modular construction.

9.3 Summary

The main conclusions of the study have been summarised in this chapter, which also emphasises how the modularity methodology advances efficiency, sustainability, and adaptability in the construction sector. While offering evidence-based insights into their adoption, the conclusions highlight the benefits of modular approaches in tackling industry challenges. The suggestions made here provide doable tactics for improving implementation through industry cooperation, training, innovation, and policy support. When combined, these components offer a methodical way to apply modularity methodology in real-world situations. Thus, the chapter ensures both academic contribution and practical relevance for the construction industry by bridging research findings with practical recommendations.

MIX
Papier aus verantwortungsvollen Quellen
Paper from responsible sources
FSC® C105338

If you have any concerns about our products,
you can contact us on
ProductSafety@springernature.com

In case Publisher is established outside the EU,
the EU authorized representative is:
Springer Nature Customer Service Center GmbH
Europaplatz 3, 69115 Heidelberg, Germany

Printed by Libri Plureos GmbH
in Hamburg, Germany